Embedded Software Timing

Peter Gliwa

Embedded Software Timing

Methodology, Analysis and Practical Tips with a Focus on Automotive

 Springer

Peter Gliwa
Gliwa GmbH
Weilheim, Germany

ISBN 978-3-030-64146-7   ISBN 978-3-030-64144-3   (eBook)
https://doi.org/10.1007/978-3-030-64144-3

Translated and Extended from the German Edition P. Gliwa "Embedded Software Timing" © 2020
Springer Fachmedien Wiesbaden GmbH

This Springer imprint is published by the registered company Springer Nature Switzerland AG.
The registered company address is: Gewerbestrasse 11, 6330 Cham, Switzerland

Preface

Embedded software makes up only a comparatively small part of the larger topic of computer science. Within this, the topic of "timing" focuses only on one specific aspect. So, is the topic of "Embedded Software Timing" one that is only relevant to a few experts?

At this very moment, billions of embedded systems are in use worldwide. Embedded software is running on every one of those devices with each system having its own set of timing requirements. If those timing requirements are not met due to a software error, the range of possible outcomes varies enormously. Depending on the product and situation, this may range from not being noticed, to being an annoyance for the user, to costing lives.

A good understanding for the timing challenges of embedded systems enables the development of better, more reliable embedded software. In addition, it is not only safety and reliability that can be improved. There are also considerable cost savings to be had across the entire development life cycle. These are not purely theoretical, as the practical examples in Chapter 6 highlight. The potential for cost savings extends across the various phases of development:

- Early consideration for the issue of timing in the design of an embedded system and its software helps decisively in increasing development efficiency and prevents timing problems from arising in the first place.

    See, among others, Sections 3.3, 6.2, and 8.1 and Chapter 9.
- Timing analysis can save time and money if the correct timing analysis technique for the given application is used. Chapter 5 provides an overview of the different techniques. Each has its own phases that describe its functional principle and workflow, highlighting use cases and limitations. In addition, an interview with one or two experts in the respective domain completes these descriptions. This livens up the topic and provides some objectivity. If the milk has already spilled—that is, if a project is already facing acute problems—troubleshooting often resembles the search for a needle in a haystack, especially in the case of timing problems. Here, too, the use of the optimal timing analysis technique delivers decisive advantages.
- Automated tests help to save costs: this is a truism. Unfortunately, existing testing all too often lacks explicit timing-related tests and focuses only on functional aspects. Section 9.6 provides recommendations in the form of concrete measures

to counteract this by ensuring embedded software timing can be verified in a well-structured manner.

- If a project reaches the point where the processor is permanently or sporadically overloaded, measures must be taken to relieve the load. This often occurs when deadlines are approaching, and, therefore, a task force is formed to relieve the situation. Chapter 8 offers knowledge that can serve as a valuable basis for such task forces. Section 8.4, which rounds off the chapter, can be used as a starting point.

The focus throughout the book is to provide a close proximity to practice. Theory is always illustrated with examples, and there are plenty of concrete tips for design, implementation, debugging, and verification.

The chapters are structured in such a way that they can be read most easily in the given order. Nevertheless, while writing this book, an attempt has been made to give each chapter a certain degree of independence, so that the reader is not lost when it is used to look something up or when reading selectively.

I would be grateful for any suggestions, criticism, and hints regarding mistakes and also welcome contact for direct professional discussion.

With that, I wish you lots of enjoyment and technical insight while reading.

Weilheim, Germany   Peter Gliwa
May 2020
peter.gliwa@gliwa.com

Acknowledgments

Time is the theme and focus of this book. To organize it in such a way that no major timing problems arose when writing over 300 pages, and creating over 100 illustrations, was sometimes a challenge.

I could not have undertaken and mastered this without the active support of many dear people.

First of all, there is Nick (Dr. Nicholas Merriam), from whom I learned a lot, such as an understanding of caches, pipelines, and spinlocks. In general, a lot of my knowledge about multi-core and runtime optimization has Nick as its source and this can now be found in the book. Thanks a lot for that, Nick!

I would also like to thank all the interview partners, not only for the time they took for the interviews but also for our collaborative work in standardization committees—they would be much less fun without you.

A big thanks goes to Stuart Cording (www.cordingconsulting.com) who not only brought the English in this book to an acceptable level but also found several flaws in its contents. Many thanks!

I would like to thank Birgit Tamkus, Peter Stief, Christian Herget, Mark Russell, and Christian Wenzel-Benner very much for their reviews and suggestions. Many thanks to you all also for supporting me in many aspects of my daily work during my writing-intensive periods.

I would like to thank my publisher *Springer*—especially, Mr. Ralf Gerstner—for their pleasant, uncomplicated, and constructive cooperation.

I still remain today grateful to Hans Sarnowski of BMW for encouraging me, back in 2002, to found a company that, from the very beginning, specialized in embedded software timing. We have also achieved many successes together on the front line of timing problems—and each and every one of them was a lot of fun.

Finally, I would like to express my greatest thanks to my wife, Priscilla, without whom I could not have written the book. Especially in the spring of 2020—and despite the Corona crisis—she freed up time for me, took care of our four children, and made it possible for me to spend many weekends and holidays in peace and quiet at the office. Many thanks for that!

Contents

General Basics

1

A prerequisite for analysis and optimization of the timing of embedded software is a basic understanding for embedded software development and operating systems. This chapter has two goals. The first is to cover or summarize important basics. The second is to establish references, where relevant, to the topic of timing. Thus, this chapter is not exclusively aimed at those who want to learn or refresh the basics. Because of this approach, even experienced software developers will appreciate the introduction of timing-related aspects to subjects that they are already familiar with.

1.1 Real-Time

If you ask a developer of desktop software or web applications what is meant by real time, you may get the answer that real time means 'very fast' or 'with very little delay'.

While this is certainly not wrong for most real-time systems, it does not get to the heart of the matter. Real time in the environment of embedded software should rather be understood in the sense of 'in time'. There are requirements with respect to time, termed 'timing requirements', which must be met. With *hard* real time, the compliance with the timing requirements must be guaranteed under all circumstances. For *soft* real time, it is sufficient if the timing requirements are not violated too often, as long as they can be statistically guaranteed. Which statistical parameter should be used to ascertain if those soft real-time requirements are met is not universally defined. If necessary, a separate definition must be found for each given situation and project.

© The Author(s), under exclusive license to Springer Nature Switzerland AG 2021
P. Gliwa, *Embedded Software Timing*,
https://doi.org/10.1007/978-3-030-64144-3_1

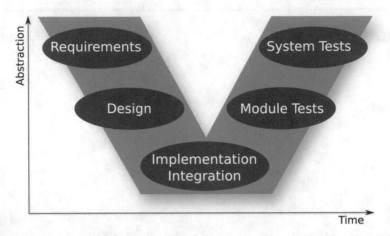

Figure 1 V-model as used in software development

1.2   Phase Driven Process Model: The V-Model

The V-model describes a concept to approach software development. It has been used for decades in the automotive industry and is usually also available—at least in the background—when newer concepts, such as Scrum, are being developed. Like so many technical developments it has its origin in the military sector. Later, it was transferred to the civilian sector and was adapted to new development requirements in the versions V-Model 97 and V-Model XT[1].

The 'V' of the V-model represents the idealized course of development in a coordinate system with two axes. The horizontal axis is a time axis with the left side marking the project start. The vertical axis describes the level of abstraction, from 'detailed' at the bottom to 'abstract' at the top (Figure 1). A project should start at a high level of abstraction with a collection of user or customer requirements for the product. This is followed by the basic design of the product at system level. Over the course of the project the design is then broken down, refined, and improved. Additional, more detailed requirements may also emerge later. Once the design phase is complete, the implementation phase begins. With respect to a software project, this corresponds to the coding. Individual components are brought together at the integration phase, and this is followed by the verification, checking the fulfillment of the requirements at the different levels of abstraction. The final phase of validation takes place at the highest level of abstraction by ensuring that the user or customer requirements are met.

If a requirement is not fulfilled, the cause of the deviation must be eliminated. The cause is inevitably somewhere on the V between the requirement and its verification. As a result, all subsequent dependent steps must also be corrected, adapted, or at least repeated.

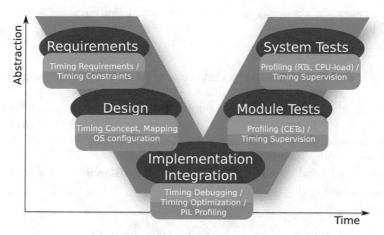

Figure 2 Software timing-related activities applied to the V-model

It is clear that the effort and associated cost to resolve an issue will grow depending on how late that issue is discovered. This reads like a truism, but it is astonishing how many projects completely neglect embedded software timing. Far too often, runtime problems are investigated in a late project phase with lots of hectic, high costs and risk, only for them to be temporarily corrected or mitigated.

1.2.1 The V-Model in Connection with Timing

Practically every software developer in the automotive sector knows the V-model as shown in Section 1.2. When using the V-model, the *functional* aspects are usually the focus. Now, what does it look like when the topic of *timing* comes into play? In principle nothing changes. The core concepts behind the model apply equally to timing. Figure 2 shows this in more detail and provides timing related examples at the different phases of the V-model.

Chapter 9 deals in detail with how timing analysis can be systematically integrated into the development process.

1.3 Build Process: From the Model to the Executable

There is an analogy between the left branch of the V-model and the process that turns source code into executable machine code, the *build process*. Both start at a comparatively high level of abstraction and, over time, get closer and closer to the executing hardware, the processor.

This section describes how to turn source code into executable machine code and which files, tools, and translation steps are relevant. The basics covered in this section are 'only' indirectly related to timing. But without an understanding of how, for example, a compiler basically works, code optimization at the code level is difficult.

1.3.1 Model-Based Software Development and Code Generation

It is probably true to say that most of the software operating in a car is model-based. This means that the source code is not written by hand but is instead generated by code-generation tools such as Embedded Coder, Targetlink or ASCET. Prior to this, the functionality—usually control loops, digital filters, or state machines—was defined using graphical modeling tools such as MATLAB/Simulink or ASCET and stored as a 'model'.

1.3.2 C Preprocessor

Listing 1 shows a simple—in this case manually coded—program. This program will be used as the basis to illustrate the path from source code to the executable. The code for the included header myTypes.h can be seen in Listing 2.

Listing 1 Simple example application (file main.c)

```
1  #include "myTypes.h"
2  #define INIT_VAL     (42)
3
4  uint32_t GetSomeValue(void)
5  {
6      static uint32_t someValue = INIT_VAL;
7      return someValue++;
8  }
9
10 void main(void)
11 {
12     volatile uint32_t a;
13     while (1) {
14         a = GetSomeValue();
15     }
16 }
```

Listing 2 Header myTypes.h as included by main.c

```
1  #ifndef _MY_TYPES_H_
2  #define _MY_TYPES_H_
3  typedef unsigned int     uint32_t;
4  #endif /* _MY_TYPES_H_ */
```

The keyword **volatile** in line 12 of Listing 1 causes the compiler to make each access to the affected variable explicit in memory, rather than allowing the value to be kept in a register for subsequent accesses. This is necessary, for example, if the contents of the affected memory location can be written to by hardware peripherals.

The timer register of a hardware timer is a good example of this. In Listing 1, `volatile` is used to prevent the compiler from 'optimizing away' the code, i.e. from determining that the variable a is never used 'meaningfully' and therefore removing all access to it.

Figure 3 shows what steps are taken on the way from source code to executable, and what intermediate formats and additional files are involved. Optional data flows, files, and tools are shown in gray.

In the first step, the compiler preprocessor parses the code and resolves all macros (`#define`), reads in all included files (`#include`), removes inactive code from conditional compilation (`#ifdef` (...) `#endif`) and calculates all values that are already calculable at this point (400 * 5 / 1000 → 2). All statements that begin with a '#' are preprocessor statements. In fact, the preprocessor performs a number of other tasks, but the examples given here are sufficient to illustrate the principle.

Hint Most compilers support the command line option -E, which causes the compiler to abort after the preprocessor stage and output the 'preprocessed' code on stdout. This can be very useful for debugging issues that relate to the preprocessor.

This output is also very useful for reporting compiler problems to the compiler vendor. If the output is redirected to a file (here the file extension .i has become common), this file can be passed to the compiler for compilation without requiring any other files. The compiler vendor can then reproduce the problem without needing access to all the header files used.

Listing 3 shows the preprocessor output as redirected to a file main.i for the source file main.c.

Listing 3 Preprocessor output (main.i) for main.c

```
1  #line 1 "main.c"
2  #line 1 "myTypes.h"
3  typedef unsigned int      uint32_t;
4  #line 2 "main.c"
5
6  uint32_t GetSomeValue(void)
7  {
8      static uint32_t someValue = (42);
9      return someValue++;
10 }
11
12 void main(void)
13 {
14     volatile uint32_t a;
15     while (1) {
16         a = GetSomeValue();
17     }
18 }
```

The #line (...) statements allow the compiler to later assign each line of the file to its original position in its original C source file. This becomes relevant when the compiler reports errors or warnings. The displayed line number of an error or warning always reflects the corresponding line in the original source file.

1.3.3   C Compiler

The output of the preprocessor is transferred to the compiler, which generates processor-specific machine code, i.e. a file with machine instructions corresponding to the C code. The memory addresses of the functions, variables, jump addresses, etc., are not yet defined at this point so they are stored *symbolically*.

The output of the compiler—in this case the TASKING compiler for the Infineon AURIX processor—can be seen in Listing 4. Since this code acts as an input for the next stage, the assembler, it is also called assembler code.

Listing 4 Compiler output (assembler code) main.src

```
1       ; source       'main.c'
2       .align         2
3       .global        GetSomeValue
4
5    ; Function GetSomeValue
6    GetSomeValue:      .type      func
7       ld.w           d2,_999001_someValue
8       add            d15,d2,#1
9       st.w           _999001_someValue,d15
10      ret
11
12      .align         2
13      .global        main
14   ; Function main
15   main:      .type      func
16      sub.a          a10,#8
17   .L3:
18      call           GetSomeValue
19      st.w           [a10],d2
20      j              .L3
```

1.3.4   Code Optimization by the Compiler

When translating source code into machine code, a compiler can perform a variety of optimizations. Many of these optimizations reduce memory requirements and, at the same time, deliver faster code. Some optimizations, however, improve one aspect at the expense of another. Here the developer must decide which aspect is more important.

The actual benefit of an optimization is often difficult to estimate in advance. During software development, the resultant output must be carefully checked. This is best done by: (a) comparing the resultant machine code for the different compiler settings and, (b) performing comparative measurements. Even experts still remain surprised by the results of such analyses. Section 8.3 deals with this topic in detail.

1.3.5   Assembler

The assembler translates the textual machine instructions of the assembler code into their binary equivalents. Thus, the output of the assembler is no longer easily readable by humans and is not shown here.

The assembler file (usually with file extension .src or .s) is assembled to an *object file*. This is often simply named *object*. As before in assembler code, the memory addresses of functions, variables, jump addresses, etc., are not yet defined in object code and remain exclusively symbolic.

1.3.6   Linker

The linker assembles all the objects passed into an almost finished program; only the concrete addresses remain absent. In our example, a single object file, namely main.o, is passed. A few more objects are implicitly added such as cstart.o, which ensures that some required basic initialization occurs before the main() function is executed. This includes the initialization of the memory interface, setting the stack pointer to the beginning of the stack, and the initialization of variables.

Additionally, function libraries can also be passed to the linker that typically come with a file extension .a or .lib. Function libraries are, in practice, nothing more than collections of objects. As shown in Figure 3 it is the *archiver* that packs the selected objects into archives. This is similar to a compression program (e.g. ZIP) or a tarball generator.

Another task of the linker is to resolve all referenced symbols. Let us assume the main function from the example calls another function SomeOtherFunction that was previously made known by an external declaration. This *forward-declaration* may look like this:

int SomeOtherFunction(int someParam);

Should this function be implemented in a source file *other than* main.c, the linker remembers the symbol SomeOtherFunction as one that is referenced but not yet defined (i.e. unresolved). In all other objects passed to the linker, the linker now searches for the symbol SomeOtherFunction. If it finds a definition, that is, an implementation of the function, the reference to the symbol is resolved. After all objects for resolving references have been searched, any function libraries passed to the linker are used to resolve the remaining references.

If the search for a symbol remains unsuccessful, the linker reports an error, typically 'unresolved external <*symbolname*>'.

If a symbol is defined in more than one object, the linker also reports an error, in this case 'redefinition of symbol <*symbolname*>'.

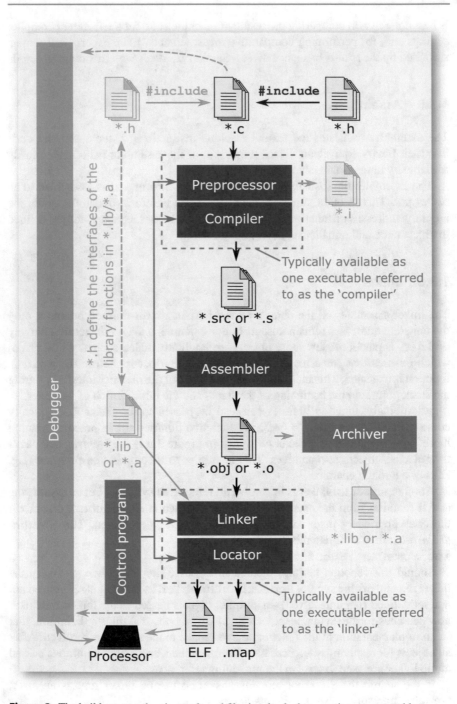

Figure 3 The build process showing tools and files involved when creating an executable

Should a symbol be defined in an object and in one or more function libraries, the linker ignores the definitions of the symbol in the function libraries and reports neither a warning, nor an error.

The order in which function libraries are passed to the linker determines the search order. Once a symbol has been resolved, all subsequent definitions are ignored and not 'linked'.

1.3.7 Locator

The vast majority of tool manufacturers combine linker and locator into a single tool that is simply named *linker*. The role of the locator is derived from its name as it 'locates' all symbols in the available memory regions. Thus, the memory addresses for each individual symbol are determined.

The output from the locator is the executable in a format that either includes or excludes symbol information. For convenient debugging of the software, amongst other things, this information is required. For example, when displaying the contents of variables, the symbol information allows you to simply specify the name of the desired variable. As a result, there is no need to determine actual memory addresses.

Typical output formats for the executable *without* symbol information are Intel HEX files (`*.hex`) or Motorola S-Records (`*.s19`). Most common for the output of the executable *with* symbol information is the ELF format (`*.elf`). ELF stands for 'Executable and Linking Format'.

Besides the executable, a *linker map*, also called *mapfile*, can be created. This file contains a list of all symbols and their memory address, to name just a few details.

1.3.8 Linker Script

The *linker script* (or *linker control file*) also plays a very important role. Strictly speaking, it should be called the 'locator script' or 'locator control file' but, as mentioned earlier, most vendors combine the locator into the linker.

Listing 5 shows an excerpt of the linker script of an 8-bit microcontroller, the Microchip AVR ATmega32, that has 32 KByte flash, 2 KByte RAM, and 1 KByte EEPROM.

The linker script tells the locator how to distribute the symbols across the different memory regions of the microcontroller. This is usually done as follows. First, in the C or assembler source code, all symbols are assigned to a specific *section* or, more precisely, a specific *input section*. This assignment will be made implicitly if the programmer has not made them explicitly. The following section names have become commonplace, representing *default sections*.

.text   Program code

        Example: `int GiveMe42(void){return 42;}`

.rodata   Read-only data
    Example: `const int a = 5;`

.bss   Read/write data, initialized to 0
    Example: `int a;`
    According to the C standard, uninitialized global variables must be ini-
    tialized with 0 by the startup code. Not all embedded software projects
    configure the startup code in this manner, so you should not rely on the
    fact that all non-initialized variables will actually be set to 0 during startup.

.data   Read/write data, initialized to a specific value
    Example: `int a = 5;`

.noinit   Read/write data, not initialized
    Example: `int a;`
    This example is identical to the one for `.bss`. Compiler switches usually
    control whether non-initialized variables in the code should be initialized
    with 0 or not at all.

.debug   Debug sections contain neither the code nor data of the program, instead
    storing additional information that makes debugging the software easier.

It is also an unwritten rule that section names begin with a dot.

In the step that follows, the instructions in the linker script assign all input
sections to *output sections* which, in turn, are finally mapped to the available
memory. For a better understanding, the `.text` sections in the Listing 5 have
corresponding comments.

In a classic linker script, as in Listing 5, the definitions of the available memory
are at the beginning. These are followed by the definitions of the output sections
and, with each of these definitions, the link to the inputs along with the allocation to
a memory region.

A very good description of the syntax of this linker script and the underlying
concepts can be found in the GNU linker manual [2]. Most other tool vendors have
at least adopted the concepts of the GNU linker (ld) for their linkers, often copying
the entire syntax of the linker script.

Listing 5 GNU ld linker script (excerpt) for the Microchip AVR ATmega32

```
1  MEMORY {
2      intflash (rx)    : ORIGIN = 0,        LENGTH = 32K
3      intram   (rw!x)  : ORIGIN = 0x800060, LENGTH = 2K
4      eeprom   (rw!x)  : ORIGIN = 0x810000, LENGTH = 1K
5  }
6
7  SECTIONS {
8      /*============= internal flash =============*/
9      /*----- program code ----*/
10     .text : {                   /* <-- output section */
11         CANdriver.*(.text)      /* <-- input  section */
12         RTOS.*(.text)           /* <-- input  section */
13         *(.text)                /* <-- input  section */
```

```
14    } > intflash                        /* <-- memory         */
15
16    /*---- constant data ----*/
17    .rodata : {
18        *(.rodata)
19        rodata_end = .;
20    } > intflash
21
22    /*============= internal RAM =============*/
23    /*---- initialized data ----*/
24    .data : AT (rodata_end) {
25        data_start = .;
26        *(.data)
27    } > intram
28
29    __data_load_start = LOADADDR(.data);
30    __data_load_end = __data_load_start + SIZEOF(.data);
31
32    /*---- zero initialized data ----*/
33    .bss : {
34        *(.bss)
35    } > intram
36
37    /*============= internal EEPROM =============*/
38    .eeprom :
39    {
40        *(.eeprom)
41    }  > eeprom
42 }
```

It may be thought confusing that input sections, output sections, and even the memories may have the same names (see eeprom at the end of Listing 5), but this is both possible and usual.

So, what does the linker script have to do with timing? As we will see in the Sections 2.4 and 2.3, location and type of access have a significant influence on the memory access duration and also the execution time of the code accessing it. The location and type of access is determined by the linker script and, therefore, knowledge of its syntax and functionality is essential for runtime optimization. Section 8.2 goes into this topic in detail.

1.4   Summary

In this chapter it was recommended to extend existing development processes with consideration for embedded software timing. The V-model was used to show how this can be done. A closer examination of the development steps with reference to timing will follow in the further course of the book.

Furthermore, the individual steps and tools in the build process were examined in more detail. Special attention was paid to the linker script because, firstly, it plays an important role in runtime optimization and, secondly, experience shows that it is often neglected, which results in problems later during development.

Microprocessor Technology Basics

2

The microprocessor, also termed 'microcontroller' or just 'controller', is the hardware unit used to execute embedded software. Without a good understanding of the basic structure of microprocessors it is almost impossible to optimize the timing of embedded software.

This chapter covers the basics of microprocessor technology with special regard for the aspects relevant to timing. Together with a microprocessor's data sheet, this chapter forms the necessary basis for the development of efficient embedded software. General basics are, as far as possible, illustrated by concrete examples. The range of microprocessors used in these examples ranges from small, 8-bit controllers to 32-bit processors, as well as single and multi-core devices.

The remainder of the book, especially Chapters 7 and 8, presuppose an understanding for the basics of microprocessor technology.

2.1 Microprocessor Design

The structure of a microprocessor is provided in Figure 4 using the example of a dual-core processor, i.e. a microprocessor with two cores.

Shown are the two computing cores, various memories, peripheral units, and two different buses. Not all processors follow this structure. The focus here is more the definition of a general concept to aid our comprehension. It should also be noted that the term 'computing core' is often reduced simply to 'core' or CPU (**C**entral **P**rocessing **U**nit).

2.1.1 CISC vs. RISC

CISC stands for **C**omplex **I**nstruction **S**et **C**omputer and describes processors whose complex machine instructions provide comparatively high functionality.

P. Gliwa, *Embedded Software Timing*,
https://doi.org/10.1007/978-3-030-64144-3_2

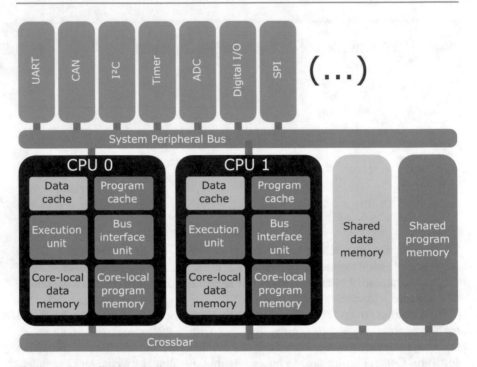

Figure 4 Block diagram of a dual-core processor

To implement such CPUs, a correspondingly complex and therefore expensive hardware is required. Furthermore, CISCs have the peculiarity that instructions take different amounts of time to execute.

RISC stands for **R**educed **I**nstruction **S**et **C**omputer. The machine instructions of such processors are simple, require few transistors, and usually take the same amount of time to execute.

2.1.2 Register

Each processor has, in its execution unit, a set of special memory cells named *registers*. Some of the commonly implemented registers are described in more detail below.

Program counter The program counter (PC) is often also termed 'instruction pointer' (IP). Each instruction in the program memory has a specific address in memory. The PC contains the address of the command that is currently being processed. Further details on the execution of the command are covered in the following Section 2.3.

Data register The data registers are used for logical operations, for calculations, and for read and write operations to and from the memories.

Accumulator RISC processors in particular have a special data register, the accumulator, which is used for most logic and arithmetic operations.

Address register Address registers are used to read data from memory, write data to memory, perform indirect jumps, or call functions indirectly. The following Section 2.3 discusses jumps and function calls in more detail.

Not all processors distinguish between address registers and data registers.

Status register The status register may also be known as a 'program status word' (PSW), 'condition code register' (CCR) or 'flag register'. It is a collection of bits where each indicates a particular state. Each bit acts like a flag and is usually abbreviated to one or two letters. The states they represent depends on the processor used with the following flags being common on most architectures.

IE, Interrupt Enable Flag Indicates whether interrupts are globally enabled ($IE = 1$) or globally disabled ($IE = 0$). To enable interrupts, further requirements must be met. Section 2.7 goes into more detail about interrupts.

IP, Interrupt Pending Flag Indicates whether an interrupt is pending ($IP = 1$) or not ($IP = 0$).

Z, Zero Flag Indicates whether the result of the last executed logical or arithmetic function was zero ($Z = 1$) or not ($Z = 0$).

C, Carry Flag Is used to flag overflows or carries for arithmetic operations as well as for logical operations. For example, if the two numbers $0xFF$ and $0xFF$ are added together on an 8-bit processor, the carry flag represents the ninth bit. The leading '1', the MSB (**m**ost **s**ignificant **b**it), of the result $0x1FE$ is in the carry flag while the remaining eight bits $0xFE$ are in the result register.

When adding with carry, the carry flag is used like a carry from a previous addition. If the Carry flag is set and the two numbers 3 and 4 are added, the result is 8.

Current CPU priority This value is found as a separate register on some processors, while on others it is part of the status register. The priority of the currently executing code determines whether an interrupt is accepted or not. Only if the interrupt has a higher priority than the currently executed code will the currently executed code be interrupted. This also requires that interrupts are enabled ($IE = 1$).

Stack pointer The stack pointer contains an address that marks the current extent of the stack used so far.

2.2   Code Execution

Section 1.3 explained how the executable machine code is generated and that this code is a collection of machine instructions. The computational core of a microprocessor is constantly processing machine instructions. These instructions are loaded sequentially from program memory (or code memory) into the execution unit, whereupon they are decoded and then executed.

The program counter (PC) has already been mentioned, and it can be thought of as pointing to the current command in the program memory. As long as there are no jump commands or commands calling a (sub-)function, the PC is increased by one memory location once the processing of a command is complete. As a result, the PC points to the next command, which in turn is loaded into the execution unit, and then decoded and executed. The program memory is primarily a sequence of machine commands.

At this point it should be mentioned that a series of machine commands without any jump or call is referred to as a *basic block*. More precisely, a basic block is a series of machine instructions whose execution always starts with the first instruction, then sequentially executes all its instructions and terminates with the execution of the last instruction. The processor does not jump into, or out of, the basic block at any other point than its first or last instruction respectively. Basic blocks play an important role, amongst other things, in static code analysis, so we will return to this topic later.

The instructions provided by a processor are described in the processor's Instruction Set Reference Manual. Knowledge of the instruction set of a processor is essential for optimizing software at the code level. Section 8.3 will cover this in detail.

How an instruction set is documented, coded, and handled will be illustrated using the example of an add instruction on the 8-bit Microchip AVR processor. Microchip AVR processors have 32 data/address registers. Their role as data register or address register depends on the instruction. Figure 5 shows an excerpt (a single page) from the instruction set reference manual for the Microchip AVR ATmega processor [3], namely the section that describes the add command with carry flag. The description in textual and operational form ($Rd \leftarrow Rd + Rr + C$) is followed by the definition of the syntax. This is the command in exactly the same notation as found in assembly code. Such an 'assembler code command' is also called a *mnemonic* . The table below the syntax displays the *opcode* of the command, i.e. the value in memory that represents the command. In this case, the six most significant bits are fixed (binary 000111) and the remaining ten bits define which registers are to be added together. The bit positions marked "d" are for register Rd, while those marked "r" are for register Rr. For example, if registers R3 and R22 are to be added and the result is to be stored in R3, the opcode looks like that shown in Listing 6. Whenever adc r3,r22 is found in the assembler code, a $0x1D63$ will appear at the corresponding position in the program memory.

5. ADC – Add with Carry

5.1. Description

Adds two registers and the contents of the C Flag and places the result in the destination register Rd.

Operation:

(i) Rd ← Rd + Rr + C

	Syntax:	Operands:	Program Counter:
(i)	ADC Rd,Rr	0 ≤ d ≤ 31, 0 ≤ r ≤ 31	PC ← PC + 1

16-bit Opcode:

0001	11rd	dddd	rrrr

5.2. Status Register (SREG) and Boolean Formula

I	T	H	S	V	N	Z	C
–	–	⇔	⇔	⇔	⇔	⇔	⇔

H $Rd3 \cdot Rr3 + Rr3 \cdot \overline{R3} + \overline{R3} \cdot Rd3$

Set if there was a carry from bit 3; cleared otherwise.

S $N \oplus V$, for signed tests.

V $Rd7 \cdot Rr7 \cdot \overline{R7} + \overline{Rd7} \cdot \overline{Rr7} \cdot R7$

Set if two's complement overflow resulted from the operation; cleared otherwise.

N R7

Set if MSB of the result is set; cleared otherwise.

Z $\overline{R7} \cdot \overline{R6} \cdot \overline{R5} \cdot \overline{R4} \cdot \overline{R3} \cdot \overline{R2} \cdot \overline{R1} \cdot \overline{R0}$

Set if the result is \$00; cleared otherwise.

C $Rd7 \cdot Rr7 + Rr7 \cdot \overline{R7} + \overline{R7} \cdot Rd7$

Set if there was carry from the MSB of the result; cleared otherwise.

R (Result) equals Rd after the operation.

Example:

```
                ; Add R1:R0 to R3:R2
add r2,r0       ; Add low byte
adc r3,r1       ; Add with carry high byte
```

Words 1 (2 bytes)

Atmel

Atmel AVR Instruction Set Manual [OTHER] 30

Atmel-0856L-AVR-Instruction-Set-Manual_Other-11/2016

Figure 5 Excerpt from the instruction set reference of the AVR ATmega processor

Listing 6 Opcode encoding for AVR "Add with Carry" command

```
adc r3,r22   ; 8bit add with carry
             ; r = 3 = 00011 bin; d = 22 = 10110 bin

             ; 1111 1100 0000 0000   Bit-position (tens)
             ; 5432 1098 7654 3210   Bit-position (ones)
             ;--------------------
```

```
 7        ; 0001 11rd dddd rrrr    Opcode definition
 8        ; 0001 1101 0110 0011    example opcode (binary)
 9                                 with r = 3 and d = 22
10        ;    1    D    6    3     same example in hex
```

Hint The comments shown in Listing 6 to explain the bit coding have proved to be very useful for programming. Whenever information is binary encoded and requires explanation, comments of this type are extremely helpful. The bit position is indicated by two comment lines: one with the tens and one with the ones. The bit position is now simply read from top to bottom. For example, 15 is for the MSB on the far left.

In addition, it is recommended that groups of four bits (a nibble) be separated from each other by means of spaces, as each nibble can also be represented by a hexadecimal number.

Under the description of the opcode in Figure 5 there follows the exact specification for which flags of the status register (which is called "SREG" for this processor) are modified and *how* they are modified.

Next is an example making use of the 'normal' add command add and then adding with carry adc. These two commands show the implementation of a 16-bit addition on an 8-bit processor.

The description of the instruction ends with the information that the opcode requires two bytes in program memory and is executed in one clock cycle.

2.3   Memory Addressing and Addressing Modes

The addressing mode describes *how* the memory is accessed. Each memory access requires the definition of the address to access as well as what should be done with the data at that address. This could range from using it to store data at the address, read from it, jump to it, call a subroutine at this address, and so on.

For runtime optimization at the code level, it is essential to know the addressing modes of the respective processor. Most processor architecture manuals (often part of the instruction reference manual) have a section that describes the available addressing modes in detail.

As the previous Section 2.2 showed, the opcode defines *what* should happen, such as 'continue program execution at address x' (a jump command), or 'load the contents of address y into working register d4'. The address *to which* some action should occur is passed as a parameter. On a 32-bit processor, the address bus has a width of 32 bits. Almost all processors are designed so that there is one byte of memory for each address. Thus $2^{32} = 4,294,967,296$ single bytes can be addressed, which corresponds to 4 gigabytes. Strictly speaking, according to the IEC [4] it should be called 4 gibibytes because the prefix giga stands for 10^9 and not for 2^{30}. In practice, however, the prefixes kibi (2^{10}), mebi (2^{20}), gibi (2^{30}), tebi (2^{40}) etc., which are based on powers of two, are hardly ever used. For this reason, we will also talk about kilobytes and megabytes in the following when referring to 2^{10} or 2^{20} bytes respectively.

But back to the 4-gigabyte address space. Most embedded systems, even those with 32-bit processors, have much smaller quantities of memory, typically ranging from a few kilobytes to a few megabytes.

If 32-bit addresses were always used this would be very inefficient as, for each memory access, the opcode as well as the full 32-bit address would have to be loaded. For this reason, all processors offer a range of addressing modes in addition to *far addressing*, the name given to the use of the full address bus width.

It is difficult to describe all existing types of addressing comprehensively and it is not useful at this point. Instead, some examples will be picked out for certain processors that differ in their implementation from the description here or have special features. Additionally, processor manufacturers have come up with a number of special addressing types that are not discussed here. For the following explanation of addressing types a fictive 16-bit processor is used. It has 64 kilobytes of program memory (flash) and 64 kilobytes of data memory (RAM). It also has eight data registers $R0 \dots R7$ and eight address registers $A0 \dots A7$. With each clock cycle, the CPU reads one word, i.e. 16 bits, from the program memory.

2.3.1   The Most Important Addressing Modes for Data Accesses

The commands in Table 1 can be classified into the following types of addressing.

Absolute addressing, far addressing   (see also LOAD Rd,Adr16 in Table 1) The address contains as many bits as the address bus is wide. Although this is the most inefficient way to access memory it is also the only one without any range restrictions. No matter where the desired memory location is located, it can be addressed. Figure 6 shows the 64 kilobytes of memory on the processor for which Table 1 provides an excerpt from the command reference. With far addressing, data can be loaded from any address.

Absolute addressing, near addressing   (see also LOAD R0,Adr12 in Table 1) The address contains fewer bits than the address bus is wide. This means that only a limited memory area can be addressed, but results in a more efficient command. According to the "Cycles" and "Memory Requirement" columns, near addressing requires half the time and half the program memory compared to far addressing. This is achieved by encoding the address into the opcode instead of loading it separately. The 12 bits allow addressing of memory locations within a four kilobyte block. Since the upper address bits are padded with zeros, this four kilobyte near area is located at the bottom of the memory, see Figure 6. The near area works in a similar way to keeping frequently used papers and documents in a drawer unit on castors under your desk. For all other documents, you will need to go to the filing cabinet with its many file folders. Although it can hold much more data, access to it is more cumbersome and inefficient.

Table 1 Excerpt from the instruction set manual of a fictive 16-bit processor

Mnemonic	Description	Cycles	Memory Requirement
LOAD Rd,Adr16	Reads a word at address $Adr16$ from the data memory and writes it to Rd $Rd \leftarrow [Adr16]$ Rd = destination register R0..R7 $Adr16$ = 16-bit address Opcode: 0100 1ddd 0111 0000	2	2-byte opcode + 2-byte address = 4 bytes
LOAD R0,Adr12	Reads a word at address $Adr12$ from the data memory and writes it to $R0$ $R0 \leftarrow [0x0FFF \,\&\, Adr12]$ $R0$ = destination register R0 $Adr12$ = lower 12 bits form the address, upper 4 bits are zero Opcode: 0101 AAAA AAAA AAAA	1	2 bytes for opcode and Adr12
LOAD Rd,@As	Reads a word from the data memory at the address specified by As and writes it to Rd $Rd \leftarrow [As]$ Rd = destination register R0..R7 As = address register A0..A7 Opcode: 0100 0ddd aaa1 0010	1	2-byte opcode
LOAD Rd,@As+	Reads a word from the data memory at the address specified by As and writes it to Rd. Afterwords, As is incremented by 2. $Rd \leftarrow [As]; As \leftarrow As + 2$ Rd = destination register R0..R7 As = address register A0..A7 Opcode: 0110 0ddd aaa1 0010	1	2-byte opcode
LOAD Rd,@A0+of8	Reads a word from the data memory at the address specified by $A0$ plus the offsets $of8$ and writes it to Rd $Rd \leftarrow [A0 + of8]$ Rd = destination register R0..R7 $of8$ = 8-bit offset 0..256 Opcode: 1011 0ddd oooo oooo	2	2-byte opcode

Register-indirect   (see also LOAD Rd,@As in Table 1) With this addressing mode the opcode does not contain any direct address information. Instead, it indicates the address register where the desired address can be found. As a result, the instruction can be executed in a single cycle and requires only one word (16 bits). However, the desired address must first be written into the address register As.

Register-indirect with post-increment   (see also LOAD Rd,@As in Table 1) This addressing mode relates to the previous one but, after the data is loaded from memory, 2 is automatically added to the contents of the address register.

Thus, the register now points to the next word in memory, hence the name 'post-increment': *incrementation* occurs *after* the access.

Addressing of this type is highly suited to accessing a block of data values stored in memory, one after the other, that must be accessed in strict sequence. This commonly occurs, for example, in the processing of arrays.

Register-indirect with offset, relative addressing   (see also LOAD Rd,@A0 of8 in Table 1) Similar to the LOAD Rd,@As command, access is indirect via an address register, but the 8-bit offset *of8* is added to this register before access. The content of the address register itself remains unaffected.

Figure 6 Data memory segmentation of the fictive processor

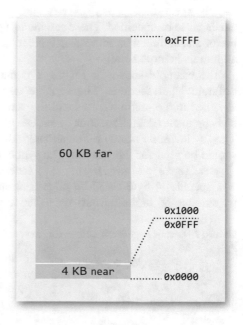

2.3.2   Addressing Modes for Jumps and Calls

For jumps and function calls, Table 1 could be extended with similar examples. Again, there are far and near ranges, direct and indirect jumps, function calls, and commands with relative addressing via offsets, usually relative to the current instruction pointer.

In addition, there are instructions whose execution is linked to a certain condition, such as being dependent on whether the zero flag is set. These are used when the compiler compiles code such as if (a!=0) {(...)}. The code in brackets is simply skipped if *a* has the value zero, which is confirmed by using the zero flag.

2.3.3   Selecting Addressing Modes

When writing C code, how do you determine which type of addressing should be used? Again, the answer depends on the processor and compiler used, so instead of a complete list, the most important mechanisms are discussed here.

2.3.3.1 Explicit Usage of Sections

One option that is practically always available is the explicit use of 'sections'. Section 1.3.8 described the basic section mechanisms and the following example is intended to make the procedure more tangible. When defining variables, they can be assigned to a specific input section by means of a special statement. These are usually **#pragma** ... or __attribute__(...) statements that allow additional attributes to be specified. These can include the definition of which input section to use or an alignment or addressing type, to name just a few. Section 8.2.2 deals with alignment in more detail.

The example shown in Listing 7 defines two variables that are assigned to different input sections. The variable myFrequentlyAccessedVar is explicitly assigned to the section .sbss and the variable myRarelyAccessedVar implicitly ends up in the (default) section .bss. See also the list of default sections on page 9. The "s" in .sbss stands for small and indicates that an efficient addressing mode should be used for this section. The syntax of this fictive example is based on the __attribute__((section(...))) statement of the HighTec GCC compiler. The a=4 specifies 4-byte or 32-bit alignment and the flags wBs specify the attributes w (writable), B (uninitialized) and s (use small addressing).

Listing 7 Example application containing near and far data accesses

```
 1  // sbss = small bss
 2  int myFrequentlyAccessedVar
 3      __attribute__((section(".sbss","a=4","f=wBs")));
 4
 5  int myRarelyAccessedVar;
 6
 7  void main (void)
 8  {
 9      volatile int a;
10      volatile int b;
11      // (...)
12      a = myFrequentlyAccessedVar; // LOAD R0,Adr12
13      b = myRarelyAccessedVar;     // LOAD Rd,Adr16
14      // (...)
15  }
```

The linker script shown in Listing 8 is valid for the fictive processor with its 4 KB near addressing area in RAM. The script places all symbols of the .sbss section at the beginning of the internal RAM (intram) and then checks that the 4 KB limit has not been exceeded (ASSERT). After that, the regularly addressed data follows in the .bss section.

Listing 8 Extract linker script with definition of the 'near' data area .sbss

```
1    (...)
2
3    /* the following output section needs to be placed
4       at physical address 0x0000 of the RAM */
5    .sbss : Adr12 /* use 12-bit near addressing mode */ {
6        __sbss_start = .;
7        *(.sbss)
8        __sbss_end = .;
9    } > intram
10
11   ASSERT( __sbss_end - __sbss_start <= 4K,
12           "error: output section sbss overflow")
13
14   .bss : {
15       *(.bss)
16   } > intram
17
18   (...)
```

2.3.3.2 Automatic Assignment to Near Section Based Upon Size

In general, very large collections of data (such as arrays) and large functions are not accessed as frequently as small data and compact functions. Based on this observation, a very simple strategy for locating the symbols in memory can be found: small symbols go in the fast memory, large symbols in the slow memory.

There is another effect. The mere fact that a larger quantity of symbols end up in the fast memories means that the faster addressing modes are used more often. It is obvious that more symbols fit into a limited memory if predominantly small symbols are assigned to it. As a result, the software as a whole works more efficiently (faster).

The advantage of this automatic assignment is its simple configuration. The compiler is told, in bytes, what a 'small symbol' is considered to be. Using the HighTec GCC AURIX compiler, the parameter -msmall-data=<size> is used for data access symbols. All data symbols (variables) with a size less than, or equal to, size are packed into the input section .sbss. The name sbss stands for small bss and now it should be clear where the 'small' comes from and what is meant by it.

The disadvantage of the automatic assignment is that it is left to chance which of the small symbols are assigned to fast memory and which are not (should there not be enough space for all the small symbols). It is very likely that some of the frequently accessed symbols are assigned to slow memory, resulting in accesses that are very slow and inefficient.

2.4   Wait States and Burst Accesses

There are a variety of different types of storage, all of which have their advantages and disadvantages. RAM is fast as well as readable and writable, but is said to be volatile as it loses its contents if not permanently powered. Flash is persistent memory (non-volatile) but access to it is relatively slow. In most cases it is so slow that access to it must be artificially slowed down. This is achieved by using 'wait states' for each access during which the processor waits for the memory to respond.

As discussed in detail in Section 2.3, each time memory is accessed, the address to be accessed must be specified. With respect to the transfer of user data, the exchange of address information can be seen as a kind of overhead (Figure 7). During the execution of code, several memory locations are very often read in sequence, especially whenever there are no jumps or function calls (keyword: basic block). The same applies to the initialization of variables with values from the flash: the values are often stored in memory one after the other.

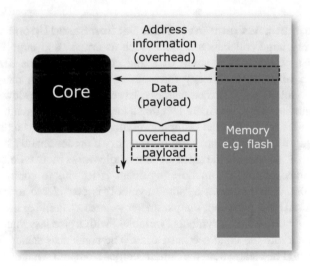

Figure 7   User data and overhead for simple read access

In both cases there would be many individual read accesses, each with significant overhead (Figure 8). To make this type of access more efficient, many memories offer *burst accesses*. These can transfer an entire range of data starting from a single address (Figure 9), significantly reducing the overhead.

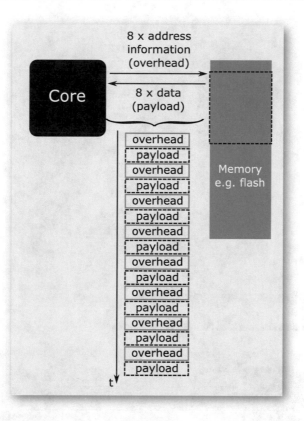

Figure 8 User data and overhead for multiple individual accesses

2.5 Cache

In a tax office, a clerk deals with the affairs of four clients in one morning. Her files are put on the desk for quick access. After all, she has to look at individual documents again and again and does not want to fetch the file from the archive for each document and then return the file back to the archive after viewing it. That would be inefficient.

This office procedure describes the concept of a *cache* very well. A comparatively small but very fast memory (desktop equates to cache) is loaded with the current contents of a much larger, but also much slower, memory (archive equates to flash or shared RAM), as in Figure 10.

With larger processors, further gradations or cache levels come into play. The example of the tax office could be extended as follows to illustrate multi-level caches. Between the desk and the archive there may also be a drawer unit on castors under the desk, as well as a filing cabinet in the office. This results in the following gradation: desk equates to level 1 cache, hanging file register equates to level 2

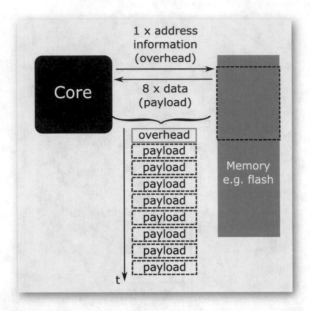

Figure 9 User data and overhead during a burst access

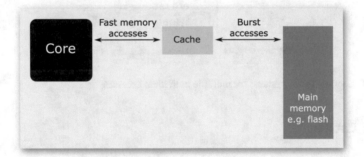

Figure 10 Principle of the cache

cache, filing cabinet equates to level 3 cache, and finally archive equals flash or shared RAM. Usually the word 'level' is not written out in full but simply replaced by an 'L'. Thus we speak of an L1 cache, L2 cache and so on.

If data or code is to be read and it is already in the cache, this is called a *cache hit*. If they are not in the cache and must first be fetched from main memory, there is a *cache miss*.

2.5.1   Cache Structure and Cache Rows

Each cache is divided into *cache lines*, each line being several dozen bytes in size. The main memory is an integer multiple larger than the cache, so the cache fits in 'n times'. When transferring data to or from the cache, an entire cache line is always transferred by burst access.

The assignment of cache lines to the memory addresses in the main memory is not freely selectable. Instead, it results from the position of the line in the cache. Figure 11 illustrates the relationship. Cache line 3, for example, can only be matched with memory areas marked with a '3'. In reality, the size ratio is more pronounced than the 1:4 ratio used in the figure and the number of cache lines is also significantly higher. Table 2 shows the parameters as they are defined for first generation Infineon AURIX processors.

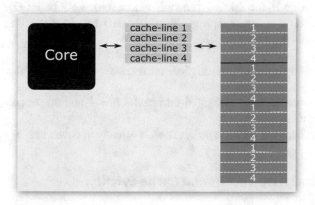

Figure 11   Cache lines and their relation to address areas in the main memory

To illustrate how the cache works, let us assume a concrete situation in which a function FunctionA has already been loaded into a cache line (Figure 12). Obviously, the function is small enough to fit completely into a cache line. Three different cases will be considered below.

Table 2   Cache parameters
of the Infineon AURIX
processors

Parameter	Performance core 1.6P	Efficiency core 1.6E
Size of flash	e.g. 4 MB	
Program cache	16 KB	8 KB
Data cache	8 KB	128 Bytes
Cache line size	256-bit (eight 32-bit words), 4-way associative (will be explained later)	
Eviction strategy	Pseudo LRU (will be explained later)	

Figure 12 Illustration using
the cache with three functions

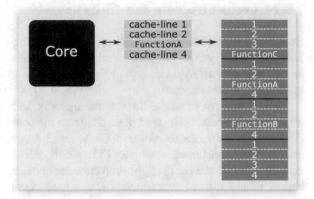

What happens if the cached function `FunctionA` now calls: (I) the function `FunctionB`; (II) the function `FunctionC`; or (III) the function `FunctionA` (i.e. recursively calls itself)?

 (I) Function `FunctionB` is loaded into cache line 3 and thus overwrites `FunctionA`.
 (II) Function `FunctionC` is loaded into cache line 4 and `FunctionA` remains in cache line 3.
(III) Nothing happens because `FunctionA` is already in cache line 3.

2.5.2 Associative Caches and Cache Eviction

The allocation of memory areas of the main memory to the cache lines, as described above, inevitably results in a certain randomness as to which cache line is loaded next for larger jumps. The contents of the corresponding line are overwritten, or 'evicted', and lost.

As a result, content that is needed again a short time later may be lost. To counteract this, *associative caches* with multiple side by side lines have been created (Figure 13). Now, when loading a cache line, it must be decided *which* of the associative lines is evicted. There are various strategies, only some of which will be mentioned here.

LRU   Least recently used; the line that has been unused longest is evicted.

Round Robin   The lines are evicted in turn.

Pseudo LRU   Approach as LRU but with a simplified logic.

Random   The line to evict is decided by chance.

Furthermore, there is the ability to 'freeze' code or data in the respective cache by using a *cache lock*. Until explicitly released, the protected content cannot be evicted by the regular cache mechanisms.

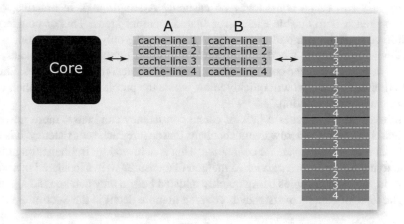

Figure 13 A two-way associative cache

2.5.3 Cache Consistency and Cache Coherency

Up until now, the processor core was the only instance that used data from the cache and, indirectly, from the main memory. However, in reality there are other protagonists that have read and write access.

These include on-chip peripherals that have direct access to the memory via DMA (Direct Memory Access). The advantage of this concept is that the peripherals can read or write large amounts of data to or from memory without the interaction of the CPU, i.e. without the CPU having to execute code.

An example is the SPI (Serial Peripheral Interface) connected to an SD card. The CPU requests the transfer of a file (it still needs to execute code to initiate the transfer process) where the actual transfer of the data within the file into the processor's memory is undertaken by the DMA.

It is not only DMA-capable peripherals that can access the memory in addition to the CPU. In a multi-core processor there are several CPUs, all of which read-from and write-to the memory. This can cause issues of *cache consistency* when the memory area in question is cache-enabled.

To illustrate the problem of cache consistency, the dual-core processor from Figure 4 is used. Suppose CPU 0 is reading data from shared data memory (shown in the figure) with the data cache turned on. This data is loaded into the data cache. Now CPU 1 changes (writes to) the shared data memory, resulting in a change to the data. The contents of shared data memory and data cache for CPU 0 are now different; they are not consistent. In and of itself, this temporary inconsistency in data is not an issue—until CPU 0 reads this same data location and acquires the (old) data value from the cache, rather than the (new) data value currently in the data memory. Issues of cache consistency are not only caused by CPU memory accesses; they can also be the result of any peripheral, such as the DMA, writing to shared memory that is also cached.

So, how can cache consistency be ensured? A simple and, in practice, often used approach is to disable caches for shared memory areas. This of course puts additional strain on computing power, but it is often the only viable solution.

Should a project be faced with sporadically occurring data inconsistencies, one way to determine if cache consistency is the cause is to deactivate the caches. Should this eliminate them, you will quickly know where the problem is and can then work to find an efficient solution.

To overcome the issues a lack of cache consistency can cause, more powerful processors implement hardware mechanisms that ensure cache consistency. It is then said that they guarantee *cache coherence*. This is achieved by implementing clever logic to invoke *write propagation*, a method that ensures writes to shared memories result in any cached copies being quickly updated before they can be read again.

The topic of data inconsistency is covered in more detail in the Section 2.9.

2.6   Pipeline

The execution of a single instruction is divided into several steps. The number of steps per instruction depends on the processor architecture and ranges from two (as with the Microchip AVR) to over 30 on some Intel processors. Typical steps include:

1: Fetch   Loading the command from memory or cache

2: Decode   Decoding, i.e. interpreting the opcode

3: Execute   Execution of the command

4: Write-back   Writing back the result (if required)

The processor clock is used to switch from one step to the next.

While one command is being decoded (level 2), the next one can already be fetched (level 1). The previous two commands are simultaneously passing through the execute (level 3) and write-back (level 4) steps. This approach allows the execution of several instructions to be run in parallel, which greatly improves the performance of the processor. The flow of instructions operates like a *pipeline*. When code is processed linearly, there are always as many instructions in the pipeline as there are stages. Figure 14 illustrates how the pipeline works.

The code shown in Figure 14 does not contain any jumps. But what happens if there is a jump and the commands immediately following the jump instruction are not executed? In this case a simple pipeline will discard all commands that are in the pipeline after the jump is detected at the decode stage. The intended program flow is thus maintained. However, the efficiency of the processor suffers because the execution rate is no longer one instruction per processor clock.

2.6.1   Branch Prediction Unit

Since program code usually has many jumps and subroutines (calls), the *Branch Prediction Unit* mechanism was developed that, in most cases, reestablishes the

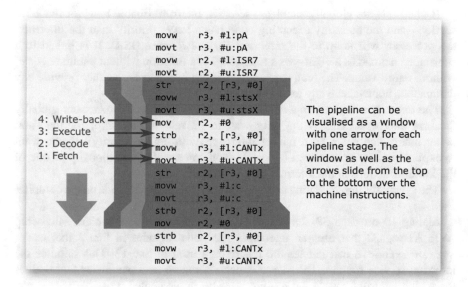

```
                        movw    r3, #1:pA
                        movt    r3, #u:pA
                        movw    r2, #1:ISR7
                        movt    r2, #u:ISR7
                        str     r2, [r3, #0]
                        movw    r3, #1:stsX
                        movt    r3, #u:stsX
4: Write-back  ─────►   mov     r2, #0                  The pipeline can be
3: Execute     ─────►   strb    r2, [r3, #0]            visualised as a window
2: Decode      ─────►   movw    r3, #1:CANTx            with one arrow for each
1: Fetch       ─────►   movt    r3, #u:CANTx            pipeline stage. The
                        str     r2, [r3, #0]            window as well as the
                        movw    r3, #1:c                arrows slide from the top
                        movt    r3, #u:c                to the bottom over the
                        strb    r2, [r3, #0]            machine instructions.
                        mov     r2, #0
                        strb    r2, [r3, #0]
                        movw    r3, #1:CANTx
                        movt    r3, #u:CANTx
```

Figure 14 Illustration of a four-stage pipeline

high rate of code execution. The branch prediction unit guesses, even before the complete decoding and execution of an instruction, which address will be accessed next. For example, it must predict whether a conditional jump will be executed or not. In more complex conditions, this prediction is not possible as it would involve the logic taking over the role of the decode and execute stages in their entirety in order to know the decision regarding the conditional jump.

The situation is similar when several (simple) jumps or function calls occur in quick succession or sequentially.

In such cases, subsequent instructions already loaded into the pipeline must be discarded, and the pipeline 'stalls'. The discarded instructions leave gaps that are simply ignored by all stages as they pass through the pipeline together with the remaining valid commands.

2.7 Interrupts

An interrupt is a kind of subroutine that is not called by the software but is instead executed as a result of a hardware event. Such events can include the receipt of a CAN message or a character via the serial interface. A level change (high to low or vice versa) at a port pin, which has been appropriately configured, is another classic interrupt source.

If an interrupt is triggered it becomes 'pending' and the interrupt pending bit in the associated interrupt control register changes from 0 to 1. If a global interrupt pending flag exists, it is also set to 1.

If interrupts are globally enabled—see "IE, Interrupt Enable Flag" in the 2.1.2 section—and the currently executing code has a lower priority than the interrupt, the processor will jump to the *Interrupt Service Routine* (ISR). It is linked to a subroutine defined as a `void-void` function, i.e. a function without parameters and without return value. The hardware as the calling instance can neither provide any parameters, nor evaluate any return value.

The entry into the interrupt service routine on almost all processors globally disables interrupts. This is the only way to prevent the interrupt routine from being interrupted again by a further interrupt (with higher priority). If you want to explicitly allow such nested interrupts, you can enable them globally as one of the first instructions in the interrupt service routine.

The syntax for implementing an interrupt service routine depends on the compiler used.

Listing 10 on page 38 shows a small executable program for the Microchip AVR ATmega32 that uses an interrupt. The code includes in line 2 the header `avr/interrupt.h` that defines the macro `ISR` used in line 13. This initiates the interrupt service routine. The macro expects a macro as a parameter that defines for which interrupt source the routine is to be implemented. Possible values can be taken from the processor-specific header. In the example, the interrupt is always triggered when the Timer 1 of the processor overflows (`TIMER1_OVF_vect`).

2.8   Traps/Exceptions

Exception handling, also known as 'exceptions' or 'traps' by some processor manufacturers, are very similar to interrupts. They generally have a higher priority than interrupts and are triggered by errors detected in hardware by the processor. Examples of such errors include a division by zero, a drop in supply voltage, destabilization of the clock generator, and violation of memory access rights detected by the MPU (Memory Protection Unit). Larger processors can often detect over one hundred possible errors that can trigger exceptions.

Analogous to the interrupt service routines, exceptions also have handling routines known as an 'exception handler' or 'trap handler'.

In many cases, exceptions help to uncover software errors during development. In other cases (power failure or destabilization of the clock generator) attempts can be made to bring the system into a safe state in the exception routine.

2.9   Data Consistency

The problem of inconsistent data will be explained using a simple example. Assume an application has two interrupts of different priority and needs to know the sum of the executions of both interrupts. A (too) simple implementation of this requirement can be seen in Listing 9. As an example, the interrupt notation of the TASKING compiler for the Infineon AURIX is used here.

Listing 9 Code to illustrate data inconsistency

```
1  unsigned int counterISR = 0;
2
3  void __interrupt(0x05) ISR_low_prio (void)
4  {
5      _enable(); // globally enable interrupts
6      counterISR++;
7      DoSomething();
8  }
9
10 void __interrupt(0x30) ISR_high_prio (void)
11 {
12     _enable(); // globally enable interrupts
13     counterISR++;
14     DoSomethingElse();
15 }
```

Apart from the fact that the overflow of the counter `counterISR` is not handled, the code has another problem.

Let's assume that 24 interrupts were counted thus far and now the low priority interrupt is triggered and the Interrupt Service Routine (ISR) `ISR_low_prio` is jumped to accordingly (Figure 15). The value 24 is loaded from memory into a register and, before the ISR completes, it is itself interrupted by the ISR of the other interrupt. Again, the value 24 is loaded from memory into a register, incremented by one to 25, and written back to memory. The interrupted ISR with low priority is then continued and the value in the register is also increased by one to 25 and written back into memory. In doing so, the value previously written by the `ISR_high_prio` is overwritten and lost.

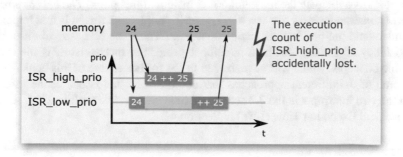

Figure 15 Example of data inconsistency

Although the code is very simple and would function in an environment without interrupts, it fails in the embedded software area. Data (in-)consistency is a central issue in the development of embedded software. Whenever code can be interrupted at any point by other code, there is always the risk of data inconsistency, even if it is not obvious at first glance.

In principle, all accesses or data manipulations that are not *atomic*, i.e. which are too long to be completed in a single CPU cycle and can therefore be interrupted, are at risk. This is especially true for data structures. As an example, consider a component that reads the first part of a structure. It is interrupted and in the interrupting routine the structure is updated. If the interrupted code is continued, it works with inconsistent data, initially with old data and then with new data.

Help is sometimes provided by the hardware. For example, when accessing timers whose width exceeds the atomic access width, processors offer special access mechanisms that ensure data consistency. Between reading the lower and upper part of a timer, the timer could overflow, which would lead to an incorrect composite timer value. The special access mechanism usually works in such a way that when reading the lower part, the hardware writes the *consistent* upper part of the timer into a separate shadow register. The upper part is now not read directly from the timer but instead from the shadow register. This ensures that the lower and upper parts are consistent with each other and the composite value is correct.

How can the problem in the example with the two interrupts be solved? A simple and, in practice, frequently encountered solution is to not allow interrupts at such critical points in execution. In this case it would be sufficient to remove line 6, i.e. the global release of the interrupts, or at least move it behind the instruction in line 7. Similarly, in most embedded software projects, code sections are enclosed by a lock and subsequent enable of the interrupts.

When operating systems (for example AUTOSAR OS) and communication layers (for example AUTOSAR COM and RTE) are used, these layers supposedly relieve you of the task of ensuring data consistency. However, problems can occur here as well, and the implementation of data consistency can require considerable resources such as runtime and memory. This topic will be discussed in detail later on, but let us anticipate the most important insight at this point. *The best assurance of data consistency is the one you don't need.* To implement this, it is first important to understand the problem of inconsistent data. Then, the best way to ensure the consistency of the data, without additional protection mechanisms, is the use of suitable concepts, which will be presented in the further course of this book.

Some of the different approaches that could resolve the issues of the example with the two interrupts in the 7.3 section include cooperative multitasking, see 3.3, and Logical Execution Time (LET) in Section 4.5.

2.10 Comparison of Desktop Processors Versus Embedded Processors

Processors for the embedded sector have always taken innovations from the high performance computing and PC (personal computing) sectors. Both areas have ubiquitous demand for more computing power in common. However, the requirements are different in their details. A desktop PC that is used, among other things, for gaming or video editing should process *on average* as many instructions per second as possible. For an embedded system that has to serve hard real-time requirements, it is the worst case scenario that is relevant.

For example, it is not an issue if the PC's mouse pointer turns into a waiting cursor for 200 ms every few hours during video editing. However, if the ignition of an airbag control unit is delayed by 200 ms, the chances of survival of the occupants in an accident are drastically reduced.

Ever smaller structures on the silicon allow ever higher clock frequencies. But this is only one reason for the dramatic increase in processor performance over the decades.

The development of more and more powerful instruction sets, longer pipelines, more elaborate branch prediction units, hierarchical caches with complex logic, and so on, contribute substantially to the permanent increase in *average* computing power. However, due to this increasing complexity, the gap between the fastest execution (best case) and the slowest execution (worst case) is also becoming wider.

Figure 16 shows the evolution of the execution time of a given function for three different processor architectures, each of which was in its prime around 20 years apart. The illustration is more about the principle than the exact proportions.

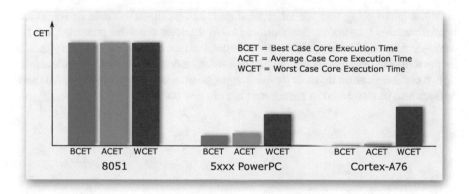

Figure 16 Evolution of execution times for identical code

The 8051, a classic among embedded processors, has neither cache nor a sophisticated pipeline. The duration of each individual instruction depends solely on the processor clock and can be looked up in the processor manual. Minimum, average, and maximum values are therefore all identical.

In the automotive sector, the 5000 PowerPC was well established for several years and used in many engine control units. It offers a cache and a significant pipeline. Accordingly, the minimum and maximum values are noticeably different when executing a function as they depend on the state of the pipeline and the state of the cache at the start of the function.

Today, in 2020, the ARM Cortex-A76 architecture is still relatively new. Average and minimum values are again much better but the extremely unlikely maximum value is comparatively high—even higher than that for the, on average, much slower PowerPC. This shows that the emphasis on maximizing average computing power can lead to the worst case scenario no longer being able to meet the requirements of a hard real-time system. This is one reason why supposedly old architectures are often used in the aviation sector. They offer the only path to guarantee the required worst case execution times under all circumstances.

2.11   Summary

After a short introduction to the basic structure of microprocessors, the instruction processing and different types of addressing were discussed using a fictitious 16-bit CPU. The examples for the different addressing types built a bridge to the topic 'linker script' of the previous chapter: symbols accessed with the same addressing type are mostly assigned to the same section.

The description of command processing and addressing types lead directly to the topics cache and pipeline, which were also covered.

During sequential instruction processing, interrupts or exceptions/traps can interrupt the program flow. If the interruption occurs at an inappropriate point, and both the interrupting and the interrupted code access the same data or resources, inconsistencies can occur. Such critical code sections must be protected by, for example, means of interrupt locks. Even better, the risk of data inconsistency should be prevented in principle. This can be achieved by using an alternative implementation, a communication layer, or by using appropriate scheduling measures. All these options will be described in more detail later in this book.

Operating Systems

3

The essential task of an embedded operating system, summarized in one sentence, is to organize the execution of the software of an embedded system. Thus, the scope of functionality is much narrower than that of desktop operating systems such as Microsoft Windows or macOS. From the perspective of such operating systems, an embedded operating system, or RTOS (**R**eal **T**ime **O**perating **S**ystem), is little more than a simple *kernel*.

There are many terms related to operating systems but they have different meanings depending on which operating system is used. This pertains specifically to terms such as task, process, thread, or exception. There is, therefore, no generally valid definition for these terms. Rather, they have to be defined in their respective context—that is, in relation to a specific operating system.

Similarly, the methodology of working with operating systems is not easily transferable from one operating system to another. Suppose a developer, who has previously developed software for Linux, is now to write software for OSEK/VDX or the AUTOSAR Classic Platform (AUTOSAR CP). They may place code that they would have packed into the endless loop of a thread under Linux into an endless loop in a non-terminating extended task under OSEK/VDX. While this works in principle, OSEK/VDX is not designed for this manner of implementing tasks and will result in several drawbacks. This example will be discussed in more detail later.

This chapter introduces different operating systems but it is limited to the aspect of *scheduling*, i.e. the rules by which different parts of (application) code, that are all competing for computing time, are organized and executed. The focus is on the question: How do I configure and use my operating system to develop software that is as resource-saving as possible, especially with regard to the timing?

© The Author(s), under exclusive license to Springer Nature Switzerland AG 2021
P. Gliwa, *Embedded Software Timing*,
https://doi.org/10.1007/978-3-030-64144-3_3

3.1   No OS: Endless-Loop Plus Interrupts

Whenever the scheduling for a system is very simple, and there is no explicit requirement for an operating system, using *no* operating system is a good alternative. I have
seen projects where just a few interrupts and a background task were implemented
using a complete AUTOSAR stack. Quite apart from the associated costs, such an
approach violates the 'keep-it-simple' principle. Less complexity practically always
entails a reduced susceptibility to errors, not to mention the fact that the operating
system itself also requires resources (runtime, stack, RAM, flash).

Now, what does 'simple scheduling' mean? Surely the periodic execution of part
of the code, the processing of another part in the background, and the presence of
some interrupts can be defined as simple scheduling. Such a configuration can be
easily realized with a periodic timer interrupt, an endless loop, and further interrupts.

3.1.1   Example Implementation with Periodic Interrupt

Listing 10 shows the implementation of a simple application with cyclic code
(DoSomePeriodicalStuff) and 'background' code (DoSomeBackgroundStuff)
and *without* recourse to an operating system. In this case it is code for the Microchip
AVR processor family. A good representative of this family is the ATmega32 upon
which the code shown can be executed.

Listing 10 Example implementation of 'infinite loop plus interrupt'

```
1  #include <avr/io.h>
2  #include <avr/interrupt.h>
3
4  void InitHardware(void)
5  {
6      DDRB = (1<<PB0); /* pin connected to LED is output pin */
7
8      /* initialize timer 1 */
9      TCCR1B = (1<<CS11) | (1<<CS10); /* prescaler = clk/64 */
10     TIMSK |= (1<<TOIE1); /* enable overflow interrupt */
11 }
12
13 ISR(TIMER1_OVF_vect) /* timer 1 overflow interrupt */
14 {
15     PORTB ^= (1<<PB0); /* toggle LED */
16     // DoSomePeriodicalStuff();
17 }
18
19 int main(void)
20 {
21     InitHardware();
22     sei(); /* globally enable interrupts */
23     while(1) {
24         // DoSomeBackgroundStuff();
25     }
26 }
```

Caution should be taken with all data exchanged or shared between `DoSomePeriodicalStuff` and `DoSomeBackgroundStuff`. Here, data consistency may have to be ensured manually, i.e. by the programmer themselves. Section 2.9 already dealt with the topic of data consistency.

3.1.2   Polling: Implementation Without Interrupts

The example shown can also be easily implemented without using a timer interrupt. An alternative approach to implementing the same application is provided in Listing 11. Data consistency problems due to interrupts can now be excluded from the outset. Instead of executing the periodic code portion in the ISR of the timer interrupt, this code portion can be placed in the same endless loop as the background code. Each time the loop is run, a query checks whether the pending flag of the timer interrupt was set as a result of a timer overflow. If this is the case, the pending flag is cleared in software and the periodic portion of code is executed. As already mentioned, this approach means that no ISR is required at all and, therefore, the background code will no longer be interrupted at any point. This permanent querying of a state is called 'polling'.

Listing 11 Implementation of the same application using polling

```
1  #include <avr/io.h>
2  #include <avr/interrupt.h>
3
4  void InitHardware(void)
5  {
6      DDRB = (1<<PB0); /* pin connected to LED is output pin */
7
8      /* initialize timer 1 */
9      TCCR1B = (1<<CS11) | (1<<CS10); /* prescaler = clk/64 */
10 }
11
12 int main(void)
13 {
14     InitHardware();
15     while(1) {
16         // DoSomeBackgroundStuff();
17         if (TIFR & (1<<TOV1)) {
18             TIFR |= (1<<TOV1); /* clear pending flag by
19                                   writing a logical 1 */
20
21             PORTB ^= (1<<PB0); /* toggle LED */
22             // DoSomePeriodicalStuff();
23         }
24     }
25 }
```

An essential difference to the version with timer overflow interrupt is that the cyclical code portion is now no longer executed with the same cyclical precision. If the overflow takes place and the timer overflow flag TOV1 is set by the hardware, it can still take some time before the function `DoSomeBackgroundStuff` ends its current call and the cyclic code section is executed. This deviation of the actual time of an event from its scheduled time, known as jitter, will be discussed later. For the time being, understand that the planned cyclic execution of `DoSomePeriodicalStuff` will be subject to jitter, the extent of which depends on the execution time of `DoSomeBackgroundStuff`. Whether the jitter becomes so large that it affects the functionality of the application must be investigated when using polling.

The trade-off between jitter and efficient data-consistency assurance will play a role again later in the context of cooperative multitasking.

Polling is always a good choice if, on the one hand, the associated delays are acceptable and, on the other hand, the waiting time is used sensibly. However, if polling is implemented as 'busy spinning'—that is, waiting in a loop exclusively for an event—one should, at a minimum, critically question the implementation.

3.1.3 Scalability

Experience shows that such approaches are expanded over time as additional periodic code parts are added. It is not uncommon to find that, after the umpteenth modification, configuration and implementation are wildly mixed, data inconsistencies occur sporadically, and the system as a whole is no longer under control. In such cases, the development team has failed to make a timely switch to the use of an operating system.

3.2   OSEK/VDX

In the 1990s, German automotive industry representatives founded a standardization committee for **O**pen **S**ystems and their interfaces for **E**lectronics in the **K**raftfahrzeug (Automobile) (OSEK), which shortly afterwards merged with the French **V**ehicle **D**istributed **E**xecutive (VDX) committee to form OSEK/VDX. In practice, OSEK is usually used for short when OSEK/VDX is actually meant.

In 2005 and 2006 parts of the OSEK/VDX standard were converted into ISO standards. The specification of the OSEK/VDX operating system [5]—from today's point of view the most important component in the OSEK/VDX standard—is available as ISO 17356-3:2005. Before we cover this in more detail we will start with a few definitions of terms.

3.2.1 Tasks

Tasks are containers for code. They have a priority and at runtime always have a defined state, as shown in the state diagram in Figure 17. If a task is in the *Running* state, its code is currently being executed. If it is in the *Ready* state, its code is ready for execution. The *Suspended* state indicates that there is no need to execute the code. These three task states—Suspended, Ready, and Running—are defined by the OSEK/VDX *Basic Conformance Class* (BCC). The OSEK/VDX *Extended Conformance Class* (ECC) comes with a fourth task state: the *Waiting* state. A task puts itself in the Waiting state by calling `WaitEvent(...)`. If the event specified in the function parameter occurs, the task's state changes from Waiting to Ready.

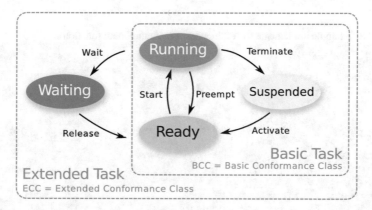

Figure 17 OSEK/VDX or AUTOSAR CP task state model

The colors for each of the states introduced by Figure 17 are used in Figure 18 to represent the state of three tasks, TASK_A, TASK_B, and TASK_C, against a time axis. This allows significantly more complex runtime situations to be visualized in a simple and easy to understand way.

To improve the clarity of such depictions, the suspended state is not visualized. If no state is visible for a task then it is currently in its suspended state.

The green arrows and labels mark the state transitions. Only the state transitions for the first occurrence of TASK_B are highlighted here.

In contrast to POSIX operating systems or even Microsoft Windows, the configuration of an OSEK/VDX based system is fixed at the time the system is created (at compile time). This means that all operating system objects, such as tasks, with all their static attributes, such as name or priority, are already known at this time. In other words, no tasks can be created or added at execution time.

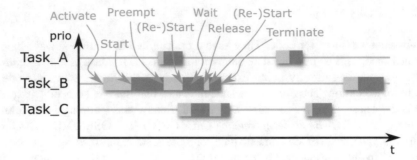

Figure 18 Task states of three tasks over time indicated by using the colors as introduced with Figure 17

Listing 12 Example implementation of three tasks including main function

```
1  TASK(TASK_A) /* Basic Task */
2  {
3    DoSomething();
4    TerminateTask(); /* must be last function call */
5  }
6
7  TASK(Task_B) /* Extended Task */
8  {
9    EventMaskType ev;
10   DoSomeStuff();
11   (void)WaitEvent( someEvent );
12   (void)GetEvent(Task_B, &ev);
13   (void)ClearEvent(ev & ( someEvent ));
14   DoSomeMoreStuff();
15   TerminateTask(); /* must be last function call */
16 }
17
18 TASK(TASK_C) /* Basic Task */
19 {
20   DoSomethingElse();
21   TerminateTask(); /* must be last function call */
22 }
23
24 int main(int argc, char *argv[])
25 {
26   DoSomeInitialization();
27   StartOS(OSDEFAULTAPPMODE); /* typically does not return */
28   return -1;
29 }
```

Listing 12 shows the implementation of two Basic Tasks and one Extended Task as shown in Figure 18.

The bodies of the tasks are defined with the macro TASK(<taskname>) and must end with the function call TerminateTask();. All OSEK/VDX operating systems map tasks to `void-void` functions by defining a macro resembling the following construct.

`#define TASK(_name_) void task_ ## _name_ (void)`

A direct call of the corresponding function by the application is not allowed. Only the operating system is allowed to call the function when the task is started.

The following options are available for activating a task at runtime:

1. **Directly via operating system service** By calling one of the operating system functions `ActivateTask(<ID of the task>)` or `ChainTask(<ID of the task>)`, the corresponding task is explicitly activated.
2. **Indirectly via Alarms** An OSEK/VDX *Alarm* is an operating system object for the cyclic activation of a task or the cyclic setting of an event.
3. **Indirectly via Schedule Tables** OSEK/VDX itself does not provide support for *Schedule Tables* as these were only introduced after the OSEK/VDX OS specification was adopted into the specification for AUTOSAR OS.

   Schedule Tables allow for an efficient and simple implementation of a recurring activation pattern for a collection of tasks. Figure 19 illustrates such a pattern for three tasks. The activation pattern in the example is repeated every 10 ms and is named a *superperiod*.

   Its period is determined by the smallest common multiple of the periods of the tasks.

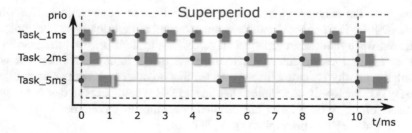

Figure 19 Three cyclic tasks and their superperiod

3.2.2 Interrupts

The code that is executed as a result of an interrupt is called an *Interrupt Service Routine* (ISR) (see also Section 2.7). OSEK/VDX supports two categories of ISRs. The first, category 1 ISRs (CAT1), run independently of the operating system (bypassing the operating system) and are therefore entirely the responsibility of the developer. The second, category 2 ISRs (CAT2), differ in that they are made known to the operating system as part of the operating system configuration. They may also call various operating system services, such as `ActivateTask(...)`, to activate a task.

3.2.3   ErrorHook

The `ErrorHook` is a function implemented by the user of the operating system and called by the operating system whenever it detects a problem at runtime. In the context of timing, failed task activation—often referred to as 'task overflow' or 'task overrun' in developer jargon—is the most important error case.

While the use of `ErrorHook` is optional by default, every project should enable it and implement meaningful responses. These should include:

* Transfer of the system into a functionally safe state (fail-safe state)
* Triggering of an error response
* Entry of the error into the diagnostic error buffer
* If tracing is in use, triggering of a suitable trigger mechanism so that the time before and after the operating system recognizes the error is visible in the trace. This enables an analysis of how the error occurred and the system's reaction to the error.

3.2.4   Basic Scheduling Strategy

What does the scheduling, that is to say the flow of control, look like when using OSEK/VDX? At runtime, tasks are activated as described and interrupts are triggered when hardware events occur. According to their *priority*, the associated task or ISR is then executed. If two tasks are in the Ready state, the task with the higher priority is started first—provided that no other task with an even higher priority is in the Ready or Running state and no ISR is being executed or is pending.

Tasks of the same priority are handled according to their order of input, i.e. according to the FIFO principle (first in, first out).

If a task is currently being executed and another task with higher priority is activated, a task switch takes place, unless the scheduler is configured to 'non-preemptive'. As the task switch takes place, the previously running task is set to the Ready state and the newly activated task starts execution.

3.2.4.1 Multiple Task Activation
The difference between BCC and ECC was explained and illustrated with Figure 17. In total, the OSEK/VDX operating system defines four conformance classes: BCC1, BCC2, ECC1, and ECC2. The type '2' conformance class indicates that tasks may be configured that have the same priority and that multiple task activation is supported. These features are not available to type '1' conformance class implementations.

Multiple task activation allows a task that is in the Ready or Running state to be activated again. With BCC1 or ECC1, this leads to a failed task activation, i.e. a runtime error. If a task is activated multiple times in a BCC2 or ECC2 configuration, the activations are buffered. This is supported up to the maximum number of activations defined in the configuration of the system. An additional task activation beyond this limit will then fail as in configuration type '1'.

In practice, multiple task activation is usually applied in the form of a bad workaround, i.e. to bypass a timing problem found during development. In most cases it would be better to eliminate the cause of the problem than deal with the additional complexity that results from multiple task activation.

3.2.4.2 Resource Protection and Priority Ceiling Protocol

During software development there are always resources, such as data, interfaces, processor peripherals, and connected hardware that have to be protected for limited periods of time. This means that, for the duration of an access by one task, it is ensured that no other task can access the same resource, even if this second task has a higher priority.

An example from everyday office life illustrates this. An employee sends a print job to the departmental printer. As it starts printing, a second print job is received, this time from the head of department. Despite the higher priority of this second job, it does not make sense to interrupt the data stream of the first job and start printing the second one immediately. The result on paper would be an unusable mix of both printouts. Therefore, the print job that has been started must first be completed.

The Priority Ceiling Protocol solves this problem by first identifying all users of a particular resource—that is, all the tasks that need access to the resource—which results in a group of tasks. The task in this group with the highest priority defines the *Ceiling Priority* of the resource. 'Ceiling' means, in this context, 'upper limit'. Ceiling priority is determined at configuration time, not during runtime.

At runtime, access to the resource is initiated by the operating system service GetResource(...). The operating system will now temporarily raise the priority of the associated task to the ceiling priority. This ensures that this task will not be interrupted by another task that may also want access to the resource. Access to the resource is then terminated with ReleaseResource(...) whereupon the task's original priority is restored.

The only tasks that can preempt the running task during the protected resource access are those with a priority higher than the ceiling priority. By definition, these cannot access this resource anyway. Blocking is limited to the duration of the access, and only to tasks with priority levels below or equal to the level of the ceiling.

In addition to the resources defined by the application there is another one called RES_SCHEDULER. This is automatically assigned the highest (task) priority. Via GetResource(RES_SCHEDULER) it is thus possible to prevent the code section that follows from being preempted by any task, until ReleaseResource(RES_SCHEDULER) is called. Interrupts however, are not blocked and remain unaffected by this protection.

In a further step, the Priority Ceiling Protocol can also be extended to category 2 (CAT2 ISR) interrupts. A priority is assigned to each CAT2 ISR which should take part in the Priority Ceiling Protocol. Everything else works as described.

3.2.4.3 Scheduling Strategy: Preemptive, Non-preemptive, and Mixed

In the Section 3.2.4 the basic scheduling strategy was described. Whenever a task with a priority higher than the currently running task is activated, a task switch occurs. Such an *immediate* switch is only undertaken during *preemptive* scheduling.

If the task's *non-preemptable* attribute is set, it cannot be immediately interrupted by higher priority tasks. In such cases, a task change can only take place if one of the following events occurs:

1. The currently running task terminates and thus moves into the Suspended state.
2. The currently running task calls the operating system service WaitEvent(...) and thus moves into the Waiting state.
3. The currently running task calls the operating system service Schedule(). If a task with a higher priority is in the Ready state, a task change takes place. Otherwise, the currently running task is simply continued.

Each task can be individually defined as preemptable or non-preemptable and a mixture of both scheduling strategies is also possible.

The naming here is a bit confusing because the scheduling strategy is defined by the configuration of the currently running task: either non-preemptive scheduling for a non-preemptable task or preemptive scheduling for a preemptable task. There is a further drawback of looking at scheduling from the perspective of the currently running task, as discussed in the following section.

3.3   Cooperative and Preemptive Multitasking

What OSEK/VDX wants to achieve with its preemptable and non-preemptable support is generally referred to as cooperative and preemptive multitasking in a real-time operating system context. The term multitasking is used here as a synonym for scheduling strategy.

The idea behind this is that there are some advantages when, instead of executing task switches as quickly as possible, i.e. preemptively, they are executed in a cooperative manner. A task in the Ready state that is waiting to be executed behaves cooperatively by waiting for the task that is currently in the Running state.

3.3.1   Illustration with Two Traces (Example 1)

For comparison, the two Figures 20 and 21 show one and the same application in the same runtime situation. The only difference is that, for the trace shown in Figure 20, `Core1_2msTask` was configured as a preemptive task and, for the trace shown in Figure 21, it was configured as a cooperative task.

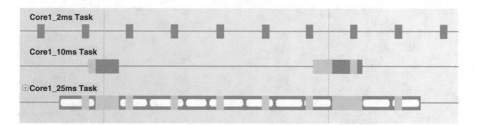

Figure 20 `Core1_2msTask` behaves preemptively with respect to the other tasks

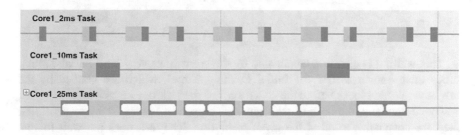

Figure 21 `Core1_2msTask` behaves cooperatively with respect to the other tasks

Unlike OSEK/VDX, the *gliwOS* operating system [6] scheduling strategy is not defined from the point of view of the currently running task but, instead, comes from the point of view of the (potentially) interrupting task. This does not, however, change the discussion of the advantages and disadvantages of cooperative and preemptive multitasking that follows in this section.

Before the advantages and disadvantages are discussed in detail, the task code and traces need to be briefly explained. In both illustrations, the ten runnables of the `Core1_25msTask` are shown as white areas within the task.

Listing 13 shows the implementation of the `Core1_25msTask`. Unlike OSEK/VDX, gliwOS does not require a call to `TerminateTask()` at the end of the task. Between the calls to the runnables, `OS_Schedule()` calls are visible, something that allows cooperative tasks of higher priority to be executed. In both traces, several cooperative task switches are visible. In Figure 20 these occur twice where switches to the `Core1_10msTask` occur. In Figure 21 these occur at the seven occasions where `Core1_2msTask` waits for a runnable of the `Core1_25msTask` to finish before it starts.

Listing 13 Code of the `Core1_2msTask`

```
 1  OS_TASK( Core1_25msTask )
 2  {
 3      Core1_25msRunnable0( );
 4      OS_Schedule( );
 5      Core1_25msRunnable1( );
 6      OS_Schedule( );
 7      Core1_25msRunnable2( );
 8      OS_Schedule( );
 9      Core1_25msRunnable3( );
10      OS_Schedule( );
11      Core1_25msRunnable4( );
12      OS_Schedule( );
13      Core1_25msRunnable5( );
14      OS_Schedule( );
15      Core1_25msRunnable6( );
16      OS_Schedule( );
17      Core1_25msRunnable7( );
18      OS_Schedule( );
19      Core1_25msRunnable8( );
20      OS_Schedule( );
21      Core1_25msRunnable9( );
22  }
```

This waiting is clearly recognizable by the sometimes very long initial waiting times (also known as IPT for Initial Pending Time), which are marked by the brightly colored blocks before the actual start of the task `Core1_2msTask`.

By comparison, the IPTs of the same task shown in Figure 20 are so short that they are not visible at all at the given resolution. The task does not wait for the runnables to complete but instead interrupts immediately. Accordingly, the white areas, i.e. the runnables, are 'cut short' at any moment in time. This is not the case with cooperative multitasking (Figure 21) where each runnable that has been started, is also completed, before any task switch can take place.

3.3.2 Stack Consumption (Example 2)

Another aspect related to the scheduling strategy are the stack requirements for the application. For this discussion we will use another example. This consists of a system with a configuration of five tasks, A to E and functions 1 to 12. The functions could be runnables of an AUTOSAR system but the following considerations also apply to non-AUTOSAR systems. The decisive factor is that the functions are called by the tasks, and require space on the stack. The precise stack requirements for each function is given in the Table 3. The two Figures 22 and 23 both reflect the same runtime situation, i.e. the activation times and (net) runtimes of functions 1 through

12, are the same in both cases. Only the scheduling strategies differ and, as a result, the stack requirements of the application.

Table 3 Stack requirement of each individual function from example 2

Function	1	2	3	4	5	6	7	8	9	10	11	12
Stack usage [Bytes]	40	50	40	50	70	60	40	20	40	30	20	40

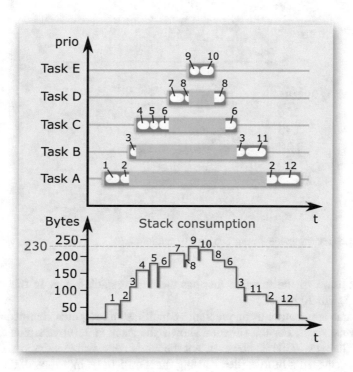

Figure 22   Stack requirement using preemptive multitasking

The second case, based on cooperative multitasking, will be considered first. All functions are executed to completion before a task switch takes place. Tasks activated in the meantime must wait as discussed, even if they have a higher priority.

So, how does this impact the stack requirements? Before Task A starts, 20 Bytes are already occupied on the stack, perhaps by the operating system itself. At the beginning of each function it 'acquires' its stack frame and releases it again at the end of the function. This explains why, at the beginning of each function's execution, the stack demand jumps up to 20 bytes plus the value specified in the table for the function, then drops back to the initial value of 20 bytes. The maximum stack usage, and thus the total stack usage of the application, is equal to the offset (20 bytes) plus

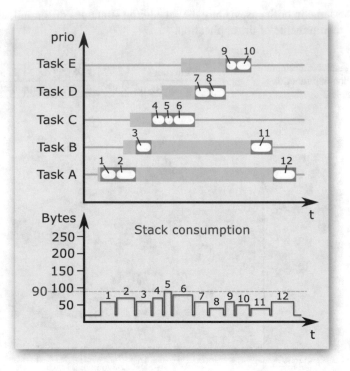

Figure 23 Stack requirement using cooperative multitasking

the stack usage of the function that has the largest stack usage. In this case it is function 5 with 70 bytes.

The same application using preemptive multitasking requires significantly more space on the stack. For the situation shown, the stack is 210 bytes plus the initial offset of 20 bytes. This is, however, not the worst conceivable situation. If Task D had been activated a little earlier, causing function 5 to be interrupted, the demand would have increased by another 10 bytes. This also applies to Task E if it were interrupted by function 7, which would have required an additional 20 bytes.

The stack requirement for preemptive multitasking is therefore not only significantly higher overall, but also much more complicated to calculate. It is also dependent on the respective runtime situation. Sometimes it is very difficult to predict if Task D and Task E could have been activated earlier or if this situation can be excluded.

In comparison, the calculation of the maximum stack requirement of the application, valid under all circumstances and for every runtime situation, is child's play for cooperative multitasking implementation. It is sufficient to know the maximum stack requirement for each function or runnable. The maximum value determined defines the stack requirement of the application.

The significantly lower stack requirement of the cooperative approach results, in many cases, in an additional effect that has a positive impact on the runtime. For projects that place the stack in a comparatively slow memory, but for which the data cache is active, significantly fewer cache misses will be observed, due to the lower stack requirement compared with a preemptive multitasking approach. In fact, the stack will often be completely cached most of the time, which significantly reduces the execution time.

Both multitasking types require that stack offset and the occurrence of interrupts must be taken into account. The same applies to functions that are called by functions from within the tasks. In reality, the stack assignment of a function over time will not result in a rectangle but rather something reminiscent of the skyline of a big city. The resultant jagged curve comes from all the subfunctions, their subfunctions, and so on, that are called.

3.3.3 Ensuring Data Consistency

The Section 2.9 has already dealt with ensuring data consistency, while Section 7.3 will review this topic again in the context of multi-core implementations. Summarized in one sentence, it is organizing access to a resource that is shared by different sections of software, so that the resource is always in a defined and correct state. This is not always easy to achieve in a system where, at any given time, the code currently being executed can be interrupted by other code, that may also be accessing the same resource.

Section 2.9 introduced the temporary disablement of interrupts as one of the possible solutions. However, this blocks *all* other parts of the code, including those that don't access the resource at all and which should perhaps be executed as soon as possible. In addition, protection formed by temporary interrupt disabling presupposes that the developer has recognized that a certain section of code needs to be protected. Such a need is not always obvious and every year software developers spend thousands of hours investigating and understanding data consistency issues that result from missing protection mechanisms.

In model-based software development, access to resources or data is specified in the modeling tool. The code generator and driver layers then ensure data consistency. Many environments, such as the AUTOSAR RTE, work in such a way that, before code is generated, it is analyzed for dependencies. This includes analyzing what parts of the code access what data, in what tasks they are running, what the priorities of the tasks are, and whether preemptive interruptions could occur. If this is determined to be the case, copies of the data are created that must be synchronized at the beginning and/or end of the affected task. The copies require additional RAM and their synchronization requires additional runtime. Today's automotive ECUs often have tens of thousands of such data items (usually termed messages) and their synchronization requires a significant portion of the available computing power.

If cooperative multitasking is used, accesses are essentially completed before a task change takes place. If configured correctly, this should also be taken into account by the aforementioned system analysis and code generation process, resulting in no security mechanisms being implemented. All the message copying and synchronization can be omitted without endangering data consistency. The resulting savings potential for runtime and RAM allocation is enormous.

3.3.4   Limitations of Cooperative Multitasking

Having read what has been said about cooperative multitasking, you could be forgiven for thinking that this would be the ideal approach. In the following section, the decisive disadvantage of cooperative multitasking will be described. Possible solutions will also be proposed that cushion this disadvantage sufficiently well, at least for the majority of embedded software projects.

The trace section shown in Figure 21 makes the disadvantage clearly visible. The start time of the Core1_2msTask is greatly delayed by the cooperative task switches. On one hand, the delay depends on the execution time of the functions in the tasks with lower priority. On the other hand, it depends on the time of task activation relative to the execution of the function currently running. The jitter (this term is explained in detail in Section 4.1.1) of Core1_2msTask is relatively high.

Whether this is a problem for the functionality of the software itself depends on the software. Most systems have quite relaxed runtime requirements that only demand that each cyclic task is executed exactly once in the time period intended for it. Whether this happens sooner or later is usually irrelevant. Assuming a BCC1 configuration, any delay in execution must not be so long that a failed task activation occurs.

How can this requirement be ensured? The answer is quite simple: the maximum runtime of the functions or runnables must be limited. However, this directly raises the next question: What is the maximum permissible limit? The answer to this question can be found in a formally correct approach, such as by using static scheduling analysis (discussed in more detail later in Section 5.9). Alternatively, a pragmatic approach can be taken by specifying a limit that is then checked using (scheduling) traces and adjusted if necessary. The maximum runtime as used here is also described as the WCET, the worst case execution time. Section 4.1.1 deals with the WCET in more detail.

Figure 24 shows a trace for a BMW Active Steering application that uses purely cooperative multitasking. Only the interrupts (top two lines of the trace) are unavoidably preemptive. However, they are implemented in such a way that they do not work directly on data used by the tasks. This can be achieved using FIFO ring buffers, for example. The result is that no protection mechanisms are required for any of the data used by the application. There is also no need to protect the FPU registers of the PowerPC used.

When cooperative multitasking was introduced in the first generation of active steering, the upper limit for the core execution time of a runnable—its allowed WCET—was set at 200 μs. This number seemed to be reasonable when considering the traces of the previous configuration. 200 μs seemed to offer a lot of room for

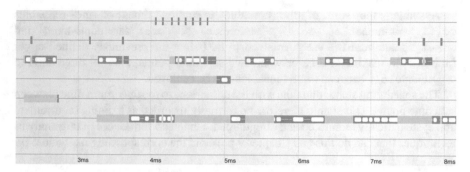

Figure 24 BMW Active Steering based on cooperative multitasking

functionality, while the delays and the jitter that resulted for the 'fastest' system task, with a cycle duration of 1 ms, were still acceptable.

However, some of the runnables had a runtime of more than 200 μs, resulting in a need to review this upper timing limit. The runnables in question were divided into several shorter runnables and, from this point on, the runtimes of all runnables were systematically recorded and checked in automated runtime measurements. Among other safeguards, the ErrorHook was implemented in such a way that, in the event of a failed task activation, an entry in the diagnostic error buffer would be made and the system would be transferred to a safe state.

This configuration, based on cooperative multitasking, and this approach to safeguarding remained practically unchanged over four generations of steering system. The vehicles in question have been in series production for years and there has never been a diagnostic error entry that indicates that the cooperative behavior of the tasks led to runtime problems. Across all generations of this solution, there have been no indication of the classic runtime problems that commonly occur, even though the processor was very heavily loaded, as the trace also shows.

3.3.5   Possible Optimization for Fewer OS_Schedule() Calls

Listing 13 provided the source code for using cooperative multitasking, while the previous section considered that the functions or runnables must not exceed a predefined execution time. Strictly speaking, it is less the execution time of any individual functions and more about the maximum amount of time between the points in time at which task switches can occur. In practice, these points in time are often referred to as 'schedule points'.

If functions exist whose maximum execution time lies demonstrably significantly below the set limit, and if the calls of these functions within the task are made directly one after the other, the schedule points in between can be omitted. When implemented in the code of Listing 13, this means that the call of OS_Schedule() can be saved at the corresponding points, providing another reduction in execution time.

Assuming the sum of the runnable execution times of the runnables `Core1_25msRunnable5`, `Core1_25msRunnable6`, `Core1_25msRunnable7` and `Core1_25msRunnable8` are guaranteed to be below the previously defined upper limit for the maximum time between two schedule points, then all three calls to `OS_Schedule()` between these runnables can be removed.

The scheduling stands and falls with the adherence to the maximum time between schedule points. This should therefore be formally recorded as a timing requirement and systematically monitored and safeguarded. Chapter 5 introduces timing analysis techniques that can do this and Chapter 9 presents the corresponding methodology.

3.3.6   Summary

A comparison of the advantages of the two scheduling strategies follows in the form of a summary.

Advantages of the Cooperative Multitasking Approach

- Results in the avoidance of a whole collection of typical real-time problems as task changes no longer take place within functions of the application or drivers.
- Correctly configured, most of the mechanisms for ensuring data consistency can be omitted, offering great potential for reducing execution time and memory requirements.
- Stack requirements are typically reduced to a fraction of that which would be required with a preemptive configuration approach.

Advantages of the Preemptive Multitasking Approach

- Short and deterministic delay times when starting higher priority tasks. This results in significantly lower jitter than with cooperative multitasking.
- Limiting the execution time of function is not required, thus making superfluous the need to split functions with longer execution times (as would be required with a cooperative approach).

3.4   POSIX

The POSIX standard is, strictly speaking, a whole collection of IEEE standards that describe, at their core, the interface between application and operating system. POSIX stands for *Portable Operating System Interface*. Basically, POSIX is used in more complex embedded systems with powerful processors. It supports the programming language C as well as the programming language Ada, the latter being mainly used in the safety relevant systems of aviation, railway, military, and nuclear power [7].

Similar to OSEK/VDX with its conformance classes BCC1, BCC2, ECC1, and ECC2, POSIX also offers different expansion stages known as 'profiles' [8]. Figure 25 illustrates this. The minimal configuration, called PSE51, defines the simplest interface implementation between the application and the operating system. As the extent of functionality of PSE52, PSE53, and finally PSE54 increases, so does the breadth or scope of the interface. Each expansion level contains the features of the previous expansion levels.

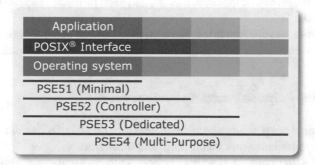

Figure 25 POSIX defines the interface between the application and the operating system

The following list summarizes the features of each version as well as explaining the usage of the terms 'process' and 'thread'.

PSE51: *Minimal* Realtime System Profile

- Application example: simple control system without user interface, file system, or mass storage.
- The system has only one processor but may have several cores.
- The application consists of a single process with one or more threads.
- The operating system provides a message-based communication interface to exchange data with POSIX operating systems on other processors.
- There is no memory management implemented (**M**emory **M**anagement **U**nit, MMU).
- There are no input and output devices.

PSE52: Realtime *Controller* System Profile

- Application example: control computer without a user interface; includes a simple file system but without mass storage
- Memory management is not necessary, but can be implemented.
- The file system uses a mass storage emulation in RAM (RAM disk) or flash. The use of mass storage is optional.

- Input and output devices are supported, but the interfaces must be non-blocking. This means that a called (interface) service must not wait internally for events and thus unduly delay the further execution of the program.

PSE53: *Dedicated* Realtime System Profile

- Application example: flight controller for the aviation sector.
- The system uses one or more processors, and each processor has its own memory management.
- Several processes are supported. Each process has one or more threads.
- The processes must be isolated from each other to limit mutual interference.

PSE54: *Multi-Purpose* Realtime System Profile

- Application example: human-machine interface (HMI) of a medical device with screen, keyboard, network connection, and mass storage.
- The application consists of elements both with and without real-time requirements.

3.4.1 Process

A process is a program that is executed both with its own data as well as data from the operating system that is required for execution. This includes status information, information about access permissions, and so on.

PSE53 and PSE54 allow several processes (programs) to be executed simultaneously. Each process is assigned a virtual memory area to which other processes have no access.

Furthermore, a process can also create new processes known as 'child processes'.

3.4.2 Thread

The processing of machine instructions runs through the program like a *thread* or, perhaps more accurately defined, like a *single* thread. The instructions are executed sequentially one after the other. Each process starts with a single thread or main thread.

If activities are to be executed in parallel, i.e. the program flow is to be split, further threads must be created. This is called *multithreading*. These threads all have access to the virtual memory of the process.

Listing 14 shows a small program that creates a thread in addition to the main thread of the process. Both access the variable counter. The program can be compiled with the GNU compiler for C and the call for its compilation is as follows:

```
g++ -std=c++11 -pthread two-threads.cpp
```

The output that the program produces is shown in Figure 26.

Listing 14 Example program `two-threads.cpp` with two threads

```cpp
#include <iostream>
#include <string.h>
#include <unistd.h>

static pthread_t tid;
static int counter = 10;
/*------------------------------------------------------------*/
void* Ping(void* arg)
{
    while (counter) {
        printf("Ping |o  | from new thread\n");
        sleep(2);
    }

    pthread_exit(static_cast<void*>(nullptr));
}
/*------------------------------------------------------------*/
int main(int argc, char* argv[])
{
    int err = pthread_create(&(tid), nullptr, &Ping, nullptr);
    if (err != 0) {
        printf( "error initializing thread: [%s]\n",
                strerror(err));
        return -1;
    }

    while (counter)
    {
        sleep(1);
        printf("Pong |  o| from main thread\n");
        counter--;
    }

    return 0;
}
```

Caution should be exercised when accessing shared data. Everything said in the 2.9 section about data consistency can also be applied to threads. For example, if several threads are to work with a data structure whose individual data fields must always be consistent with each other, a *mutex* (short for mutual exclusion object) can be used. Only the thread that has previously acquired the mutex may access the structure. After the access, the thread will release the mutex again. If a mutex is currently occupied when trying to acquire it, the process must wait until it is released again. This mechanism is very similar to the spin-locks introduced in Section 7.3.

Another way to achieve data consistency is to use the priority ceiling protocol introduced in Section 3.2.4.2, which is also available in POSIX.

```
peter@TECRA:~$ g++ -std=c++11 -pthread two-threads.cpp
peter@TECRA:~$ ./a.out
Ping |o  | from new thread
Pong |  o| from main thread
Ping |o  | from new thread
Pong |  o| from main thread
Pong |  o| from main thread
Ping |o  | from new thread
Pong |  o| from main thread
Pong |  o| from main thread
Ping |o  | from new thread
Pong |  o| from main thread
Pong |  o| from main thread
Ping |o  | from new thread
Pong |  o| from main thread
Pong |  o| from main thread
Ping |o  | from new thread
Pong |  o| from main thread
peter@TECRA:~$ █
```

Figure 26   Compiling, invocation, and output of the example under Linux

3.4.3   State Diagram of POSIX Threads

Figure 27 graphically represents the states a POSIX thread can take. The possible state transitions are also included.

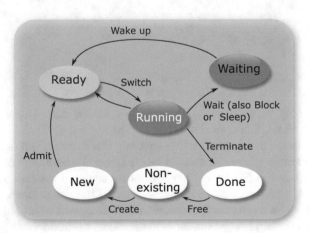

Figure 27   State diagram for POSIX threads

When a program is started, a process is created for it and its associated main thread is set to the state 'New'. After basic initialization by the operating system, the process changes to the 'Ready' state and the thread waits for its code to be executed. When execution begins, the thread is assigned the 'Running' state. There can be several reasons for the change to the 'Waiting' state. One possible reason is, as in Listing 14, the call to the function `sleep(...)` that causes the thread to sleep for the time specified by the parameter in seconds. Once the time has passed, the thread changes to the 'Ready' state. Depending on the priority and what other threads are currently waiting to be processed, the thread is returned to the 'Running' state.

The termination of a thread can have various causes. One possibility is that the function that was executed when the thread was started returns, as is the case in the function `Ping`. Alternatively, the thread may be terminated explicitly. This occurs at final point when the program ends and the **return** statement of the `main` function is reached. This terminates the program and all threads change from the state 'Done' to the final state 'Non-existing'. Only when the program is restarted is there a 'rebirth' and the life cycle begins again.

3.4.4 Scheduling Strategy

A common challenge of office life is dealing with the flood of emails that come in each day. The question that then arises is: In what order should the emails be processed? Perhaps a few with quick answers first, followed by one or two that are urgent, and then the most important ones.

The scheduler is in a very similar situation when it has several threads in the 'Ready' state. The scheduling strategy or *scheduling policy* defines the rules that determine which of the threads available for processing will be executed. The following list is only an overview of the most important scheduling policies.

Select by priority   Similar to preemptive OSEK/VDX, the thread with the highest priority is executed first.

Time slice procedure (also round-robin)   Each thread is processed for a defined period of time before the next one is given its turn.

First In, First Out (FIFO)   In this context, 'first in, first out' means that the threads are processed in the order in which they switched to the 'Ready' state.

If further information is available, such as the expected remaining time or a deadline, this can also be taken into account in the scheduling strategy. For example, the *earliest deadline first* strategy processes the thread whose deadline is closest in time. The *shortest job next* strategy is based on the idea that the tasks that can be processed quickly should be processed first, just like in the example with the emails above.

3.5   Summary

As paradoxical as it may seem, in this chapter started by dealing with systems that manage without an operating system. This was followed by a presentation of the OSEK/VDX operating system as a classic representative of an RTOS.

The section on cooperative and preemptive multitasking can certainly be seen as a recommendation to at least consider cooperative multitasking in system design. A long list of advantages of this approach contrasts only with the disadvantage of limiting the execution times of runnables or functions called directly from tasks. This is an acceptable limitation for many systems. The vast majority of data inconsistencies, as discussed in the previous chapter, can actually be prevented by the inherent implementation of cooperative multitasking.

In this chapter was then rounded off with an, admittedly very brief, introduction to the topic of POSIX. The reader will thus note the clear focus on embedded systems based upon classic real-time operating systems.

Timing Theory

4

The term 'timing theory' probably makes the topic of this chapter sound more academic than it actually is. The emphasis here is on the basics as well as explanations of terms that are relevant to everyday practice.

For example, if you were to ask three different project managers what the terms 'CPU utilization' or 'CPU load' mean, you will likely get three different answers. Different interpretations of the same terms also have an impact on timing tools too: for the same runtime situation of a certain software, three different software tools will often deliver three significantly different results for the CPU load.

Many years ago, when I first participated in a research project on timing, I was very surprised to find that there was no uniform definition or terms for many very elementary timing parameters. In response, a colleague who had undertaken his doctorate in this field, described the situation by saying that, "Academics would rather share their toothbrushes with each other than their definitions or terms."

I did not want to put up with this and created the first version of the definition of timing parameters as shown in Figure 28. Years later, I was able to provide an updated version of the definitions to the AUTOSAR standard. Since then, it has become a part of the Technical Report "Timing Analysis" [9].

What is more, current and future aspects of timing theory are also covered in this chapter. For example, the term 'Logical Execution Time (LET)' has found its way into the AUTOSAR standard, but so far only a few projects make use of it. This is despite its suitability for making systems much more predictable (deterministic) and reliable.

As far as the definition of timing parameters in POSIX operating systems is concerned, there is still some catching up to do. Although there are some clearly defined timing parameters, those required for periodic events are missing. But it is precisely these periodic events that are very important in the embedded software environment, as they play a key role in, for example, control algorithms.

© The Author(s), under exclusive license to Springer Nature Switzerland AG 2021
P. Gliwa, *Embedded Software Timing*,
https://doi.org/10.1007/978-3-030-64144-3_4

Some of the aspects dealt with below are summarized in a DIN-A1 poster. This poster is available for download as a PDF [10] as well as being part of the accompanying online book material.

4.1   Timing Parameters

In this section, a range of timing parameters will be defined. The occurrence of a task, an interrupt, or even a thread will be defined by the term 'instance', where an instance always refers to an occurrence on the time axis, and not an instance in the sense of instantiating a class in object-oriented programming.

4.1.1   RTOS Scheduling (OSEK, AUTOSAR CP, etc.) Timing Parameters

The timing parameters CET, GET, IPT, and RT covered here describe the timing of a *single* instance, such as a task or interrupt, while the parameters DT, PER, ST, and NST describe the timing *between* two such instances. Strictly speaking, the NST parameter potentially considers additional tasks and interrupts.

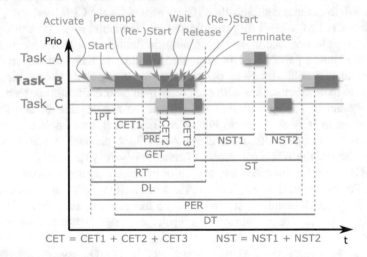

Figure 28   Definition of timing parameters based on an OSEK/VDX runtime situation. All timing parameters are related to TASK_B

In the following, the timing parameters are described in detail as they are used in OSEK and AUTOSAR CP. They can also be easily transferred to most other real-time operating systems. In the remainder of this chapter, and of this book, the abbreviations are used in some cases while, in others, the term is written in full.

CET—Core Execution Time, also net runtime

The core execution time is the central timing parameter at the code level. It indicates how much time the CPU actually spends or has spent on the execution of, for example, an interrupt, a task, a runnable, a function, or a code fragment such as a loop body. Time that the CPU spent handling any interruptions, such as those caused by interrupts or higher-priority tasks, are not included in this net runtime measurement. Such interruptions are deducted when determining the CET. See also Section 5.5.1.2.

A use case for CET would be runtime optimization at code level. If, for example, the runtime of a function is to be shortened, while of course retaining its functionality, then CET is the relevant variable to review. Whether any interruptions occur during execution, how many may have occurred, and how long they took to execute is of no interest and should be explicitly hidden in the evaluation.

If all CETs for a certain period of time are summed and compared in relation to the duration of this period of time, the result is the CPU utilization or CPU load. Section 4.3 goes into more detail on CPU load.

A special variant of CET is WCET, the **Worst-Case Execution Time**. It is a matter of definition what exactly is meant by 'worst case'. In an academic environment and also in static code analysis (see Section 5.3) this is understood to be the highest possible value of the CET under all circumstances. In practical applications, the realistic maximum value of the CET is often termed WCET.

DT—Delta Time

The delta time describes the time between the start of an instance and the start of a subsequent instance, i.e. the time difference between two consecutive start times of one and the same task.

Control algorithms or, more precisely, digital control loops, require the cyclical sampling of input values, their processing, followed by a cyclical output for further processing. The differential equation of the control loop is based upon a specific period Δt that is usually defined as a constant, for example ten milliseconds. In reality, however, the intended cycle time will not always be met. Other parts of the code, such as interrupts or higher priority tasks, cause delays that can result in the control algorithm not being called at the time intended.

The delta time reflects the *actual* cycle time. The extent to which deviations from this intended cycle time are permissible must be specified by the function developer and, if necessary, determined by stability analysis of their loop controller. In addition, it is possible to incorporate the delta time measured at runtime into the control algorithm instead of using a constant value.

PER—PERiod

In scheduling, the period describes the time difference between two successive activations of the same task. An activation marks the time at which a task has been planned to start. The period is usually important if the task is a cyclical task, that is, if the task is to be activated in a fixed, regular time pattern. The configured period, i.e. the desired cycle time, is identified in the following examples by the index 0: PER_0.

If the activation time for interrupts are known, i.e. the times at which the pending bit of the interrupt source jumps from 'not pending' to 'pending', PER can also be applied to interrupts.

Offset

The offset plays an important role in conjunction with periodic events. It describes the time offset of a periodic event, for example the activation time of a cyclical task, from an imaginary zero line. By means of the offset, the temporal positions of several periodic tasks with respect to each other can be defined. When creating periodic tasks in the operating system configuration, period and offset are set for these tasks. Section 8.1.6 describes how to optimize a system through appropriate selection of offsets.

JIT—JITter

The deviation of the actual to the desired cycle time described in the definition of the delta time above is described by means of the jitter or, more precisely, periodic jitter. The jitter JIT defines the difference between the delta time DT and the desired cycle time PER_0 in relation to the desired cycle time:

$$JIT = \frac{DT - PER_0}{PER_0} = \frac{DT}{PER_0} - 1 \tag{1}$$

If the delta time is smaller than the desired cycle time, i.e. the event under consideration comes too early, the jitter is negative. Figure 29 illustrates delta time and jitter using a section of a trace. The specified timing parameters refer to the task Core1_2msTask with a configured period of $PER_0 = 2$ ms. The vertical gray bars clearly show that the activation times are exactly two milliseconds apart. The start times of the task, on the other hand, fluctuate considerably, which ultimately leads to a jitter of over 40% in some cases. The orange values Δ_{max} and Δ_{min} are used and explained below.

The reason for the rather high jitter in this case is the use of cooperative scheduling. Before a task switch to a waiting task of higher priority can occur, the operating system allows the currently executing runnable to complete its work. The runnables are visible as white ovals in the task Core1_25msTask. Section 3.3 dealt with cooperative and preemptive multitasking in detail.

With reference to the delta time, the jitter described here is implicitly coupled to the start times of the tasks and thus becomes a 'start-to-start' jitter. In principle, a jitter can be defined for any periodic events. This is useful, for example, for the cyclical sampling points of the input data to a control algorithm (sample jitter) or the end points of the processing of cyclical runnables or functions (end-to-end jitter).

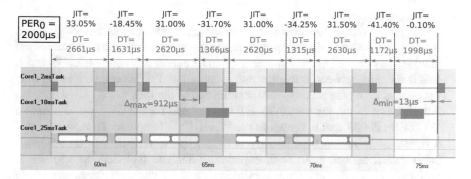

Figure 29 Illustration of Delta Time (DT) and Jitter (JIT)

J—Absolute Jitter

In addition to the aforementioned definition of jitter, there is another definition that we refer to in this book as *absolute jitter* for distinction. Absolute jitter is particularly important in static scheduling analysis (see Section 5.9.1). It refers to the nominal timing of an event relative to the actual timing. Assuming a task with the desired cycle time PER_0 exists. This results in the nominal start times of each instance of the task on the time axis. However, the task actually starts at a time Δ *after* this time. The absolute jitter is now calculated from the maximum and minimum Δ:

$$J = \Delta_{max} - \Delta_{min} \qquad (2)$$

The absolute jitter is therefore always a positive time. Table 4 illustrates the calculation of absolute jitter based on the trace shown in Figure 29.

Table 4 Calculation of the absolute jitter J for Core1_2msTask in Figure 29

Instance	DT [μs]	Start time [μs] related to first instance		Δ[μs]	Comment	Absolute Jitter
		Actual	Nominal			
2	2661	2661	2000	661		
3	1631	4292	4000	292		
4	2620	6912	6000	912	Δ_{max}	
5	1366	8278	8000	278		$J = \Delta_{max} - \Delta_{min}$
6	2620	10898	10,000	898		$= 912\,\mu s - 13\,\mu s$
7	1315	12213	12,000	213		$= 899\,\mu s$
8	2630	14843	14,000	843		
9	1172	16015	16,000	15		
10	1998	18013	18,000	13	Δ_{min}	

RT—Response Time

The response time is the most important timing parameter in scheduling theory. It tells you how much time elapses between the moment when a task or interrupt needs to be executed and the moment when its execution is complete. For tasks, this is the time between activation and termination. For interrupts, it is the time between entering the pending state and the end of the ISR.

Additionally, in the same way WCET was defined for the CET, we also have a WCRT for the RT that is known as the **W**orst-**C**ase **R**esponse **T**ime. Again, it is a matter of definition what exactly is meant by 'worst case'. In an academic environment as well as in static scheduling analysis (see Section 5.9) this is understood to mean the highest possible value of RT under all circumstances. In practical applications it is typically the realistic maximum value of the response time that is often referred to as WCRT.

DL—DeadLine

The deadline is nothing more than the maximum allowed response time. A deadline is therefore a specification; deadlines cannot be measured.

In the case of periodic tasks in an OSEK BCC1 setup, deadlines are implicitly set according to the period of the respective task. A task that is activated every ten milliseconds, for example, must terminate at the latest ten milliseconds after its activation so that the next instance can be activated ($DL < PER_0$).

GET—Gross Execution Time, also gross runtime

The gross execution time is the time difference between the start of a task and its termination, or between the start and end of an interrupt, a runnable, a function, or a code fragment. In contrast to CET, interruptions and preemptions are not 'deducted' from the gross execution. As a result, they increase the gross execution time if interrupts or task switches to tasks with higher priority occur.

In practice, gross execution time is often used incorrectly where response time (RT) should have been used due to a misunderstanding of the terminology. The difference between response time and gross execution time is that the response time also includes the initial delay between activation and start (see Initial Pending Time (IPT) described below). If interruptions and preemptions are relevant, the initial delay should also be included, being interpreted as a kind of interruption or preemption that occurs before the start.

IPT—Initial Pending Time, also initial delay

The initial pending time is the time a task waits for its start, i.e. the time difference between activation and start or, in the case of interrupts, between entry into the pending state and the start of the ISR.

ST—Slack Time

The slack time describes the 'gap' between the end of one instance of the observed object and the start of the next. It is irrelevant what happens inside this gap. Therefore, the slack time can only be used to a limited extent to determine the remaining headroom for the task or interrupt.

An example shall illustrate this. Let us assume that the configured period of a Task X is $PER_0 = 1$ ms and its slack time is $ST = 700\,\mu s$. This doesn't look critical yet, but let's assume that after each instance of this Task X, a Task Y with higher priority is running with a gross run time of $299\,\mu s$. If the runtime of the Task X were to extend by only $2\,\mu s$, the next task activation would fail (assuming an OSEK/VDX BCC1 setup).

NST—Net Slack Time
The net slack time is calculated from the slack time minus all CETs that both fall within the slack time period and belong to tasks or interrupts with higher priority. This may seem complicated to comprehend at first, but the definition of net slack time quickly becomes clear when you consider the idea behind it. As explained above, the slack time ST cannot be used as an indication for 'how much headroom' a task or interrupt has left for additional functionality. NST is, therefore, used precisely for this purpose. For the previous example with the two tasks X and Y, the net slack time of Task X is a value of one microsecond.
Figure 28 also illustrates the net slack time. When calculating the net slack time of TASK B, the CET of TASK A is taken into account because it has a higher priority than TASK B. The CET of TASK C with a lower priority, on the other hand, is irrelevant for the calculation. If, theoretically, the CET of TASK B were to increase by the value of the current net slack time in the displayed runtime situation, TASK A would interrupt at some point. TASK C, on the other hand, would not take effect. So the execution of the second instance of TASK C in the image would be delayed until after the second execution of TASK B.

PRE—PREemption Time, also interrupt time
Interrupt time does not play a major role in practice. It reflects the sum of all interruptions and preemptions, i.e. their duration, during the instance under consideration.

NPR—Number of PRemptions, also number of interruptions
The number of interrupts can either refer to a single instance of a task, interrupt, runnable, function, or code fragment, or to the sum of all interrupts for a given time period. In this second case it is a useful parameter for the scheduling implementation of the CPU being considered. Each interruption causes scheduling overhead, that is, runtime that is not available for the execution of application code. Therefore, when configuring a system, you should aim to get by with as few interruptions as possible.

4.1.2 Timing Parameters Related to POSIX

POSIX has defined comparatively few timing parameters and those defined are illustrated in Figure 30. This diagram corresponds to the earlier state diagram of POSIX threads (Figure 27) with the addition of the definitions of the POSIX timing parameters. In relation to *AUTOSAR Adaptive Platform*, Section 10.2.8.1 introduces further timing parameters that are mostly the same as those described in the previous Section 4.1.1.

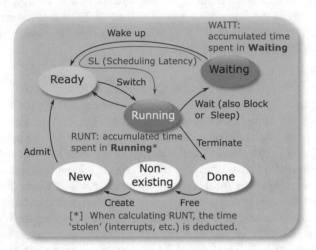

Figure 30 POSIX timing parameters

4.2   Statistical Aspects

If the timing parameters are to be described for a large number of instances, each value that occurs can be recorded and used for later analysis. Chapter 5 describes, among other things, scheduling simulation, tracing, and runtime measurement. These are all timing analysis techniques in which a large number of instances are created or observed.

Often, the processing of such a large number of measured values is cumbersome, may not be helpful, or can even be impossible to collect. If the measurement of these values is undertaken by the target system itself, there will simply not be enough space to store all the values. Therefore, measurements focus on establishing minimum, maximum, and average values.

4.2.1   Minimum and Maximum

The most important statistical parameters are the minimum and maximum values of a timing parameter. They are also extremely easy to determine for a given amount of data. With reference to a defined observation period, the largest and smallest values of all instances are simply determined for a timing parameter of an observed object (for example, a task).

A simple example illustrates this. In a trace reflecting around two minutes of execution time, 11998 instances of the task named my10ms_Task are observed. The CET was determined for all instances and the largest value observed was 1.234 ms. Thus, for the observation period to which the trace corresponds, $CET_{max} = 1.234$ ms.

4.2.2 Average

In timing analysis, the average value is understood as the unweighted arithmetic mean of all values of a defined observation period. If x is any timing parameter, the average value $\bar{x}$ for the n values in the observation period is calculated as shown in formula 3.

$$\bar{x} = \frac{1}{n} \cdot \sum_{i=1}^{n} x_i \qquad (3)$$

This is quite straightforward and clear. The situation becomes somewhat more complex when the average values are to be determined continuously, for example during an ongoing measurement or a running simulation. A total average value can only be formed *after* completion of the measurement or simulation. However, when performing measurements on a running system or simulation, the question arises of how the observation period is to be defined, i.e. over *which range* of values should the mean value be calculated?

Instead of diving into the theory of averaging at this point, the most important approaches for timing analysis are illustrated in Table 4.5. Each cell highlighted in color is an unweighted arithmetic average of all the values vertically above the cell in the row "Value x_i".

4.2.3 Histograms

Histograms visualize, for a given set of values, the distribution of the values between their minimum and maximum value. It can show whether the values are relatively evenly distributed (as indicated by bars of equal height) or, if the distribution looks more like a bell, that the minimum and maximum values rarely occur.

Figure 31 gives an example of the CET of a task where the data is derived from a trace. In the trace a total of 7031 instances of the task were observed with a minimum CET value of 122 µs and a maximum value of 176 µs.

The CET is now plotted on the x-axis in equidistant sections. The figure shows 20 such sections in this case. The y-axis reflects the number of task instances with which a CET from the respective section was observed.

Table 4.5 Comparison of different averaging methods

Index i	1	2	3	4	5	6	7	8	9	10	11	12
Value x_i	96	20	28	36	53	27	41	32	62	36	73	68
Overall average	48											
Intermediate average	45				38				60			
	45											
		34										
			36									
				39								
Moving average					38							
					41							
						43						
							51					
								60				

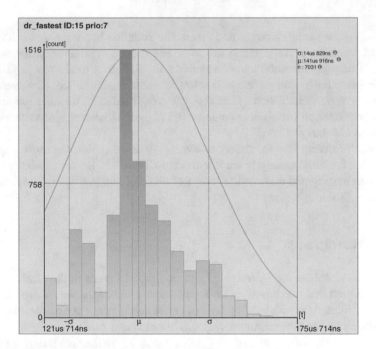

Figure 31 Histogram of the CET of a task (derived from trace data)

4.2.4 Occurrence Patterns of Irregular Events

The statistical parameters covered thus far were intended to describe a single timing parameter. For the response time of a task both minimum, maximum, and average values can be calculated, as well as using them to create histograms.

The goal with the *occurrence pattern of irregular events* is to describe a more or less sporadically recurring event. Usually this involves determining the trigger times of interrupts or the activation times of non-periodic tasks.

How such activation patterns should be interpreted is explained in Figure 32. The figure uses data from an energy management ECU where the activation pattern of the CAN receive interrupt is shown.

The x-axis shows the number n of interrupt instances starting at 2. The y-axis is a time axis. There are two curves in the diagram: an upper one in red and a lower one in green. The upper curve indicates how long the *longest* time window is in which n interrupts can occur. Respectively, the lower curve shows how long the *shortest* time window is in which n interrupts can occur. Often the axes in such diagrams are also reversed. This leads to the question: How many interruptions can be expected for a given time window?

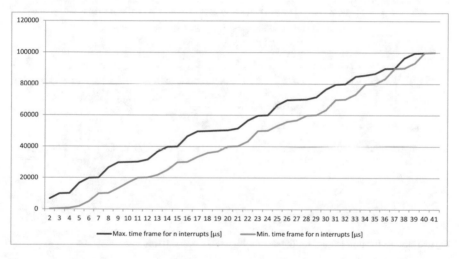

Figure 32 Activation pattern for randomly distributed interrupts

Figure 33 shows a second example. Again it is a CAN receive interrupt, this time from a steering controller. In addition to the activation pattern, the trace from which the activation pattern was obtained is shown. When looking at the pattern it is noticeable that the curves are in the form of a staircase. Every 8 interrupt instances there is a jump of around 10 ms. The reason for this is clearly visible in the trace. Every 10 ms a series of eight interrupts in short succession—a *burst* of interrupts—is triggered.

So, what are activation patterns of irregular events in timing analysis used for? They play an important role in the simulation of scheduling as well as in static scheduling analysis. Both techniques are described in detail in Chapter 5.

Whenever the mechanism by which events occur is unknown, or there is no such mechanism, activation patterns can be easily used to determine how these events

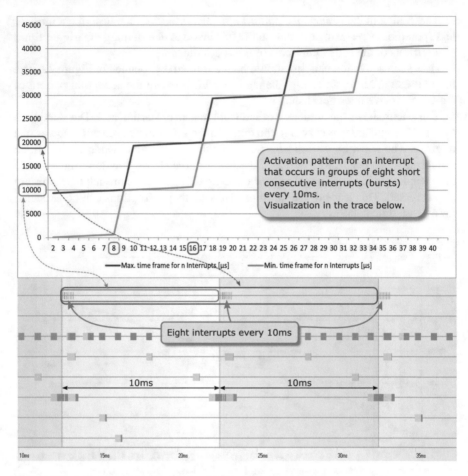

Figure 33   Activation pattern for bursts of 8 interrupts every 10 ms

occur. Such patterns can serve as input data for scheduling simulation or static scheduling analysis.

4.3   CPU Load

Specifying the CPU load for a system is an attempt to describe the state of the scheduling with a single value that reflects all the system's tasks, interrupts, delays, timing parameters, timing requirements, etc. It stands to reason that such a drastic simplification is not an easy task and must be accompanied by some compromises.

Managers especially appreciate CPU load, even more if you simplify it even further by defining them as follows: <70% green, 70% to 85% yellow and >85%

red. Developers are tempted to smile a little arrogantly at this point because they know better and have realized that the world is not quite this simple.

This section attempts to combine these two views. First, a definition of the CPU load is given, followed by a discussion of the parameters used in the calculation. It will be shown that one of these parameters can be understood as a project-specific configuration parameter. The setting of this configuration variable is done with the aim of satisfying the 'manager's view' of the CPU load, i.e. a value of 100% represents the point of overload, and any value below that is acceptable provided it can be guaranteed that it will not be exceeded under any circumstances while the system is running. Headroom for future additional functionality is a related issue and is discussed elsewhere in Section 9.7.

4.3.1 Definitions

An active CPU always executes machine instructions, even when there is no need to execute code. Idle code is code that has no functionality and is always executed when no other code is pending for execution. It is usually part of the operating system but, even when no operating system is used, a wait loop in the main function can also be considered as idle code.

The CPU load for a single CPU and for an observation period of duration t_o is calculated from the time t_e, the time the CPU spends processing code other than idle code in relation to the duration of the observation period t_o, as shown in formula 4.

$$U = \frac{t_e}{t_o} \tag{4}$$

This means that the CPU load is a number between 0 and 1. It is usually expressed as a percentage.

In Section 5.5.1.4 the CPU load as shown in formula 4 is used to implement a measurement approach that should determine the CPU load.

It is not always helpful or possible to use the execution of the idle code. In a slight variation of formula 4, a single cyclical task is the focus of attention. If it is executed cyclically with the period PER_0 and has a net run time CET, the CPU load applied for its processing is to be calculated as shown in formula 5.

$$U = \frac{CET}{PER_0} \tag{5}$$

In the approach described below, this is developed further using the CETs of tasks (or threads and processes) and interrupts. First of all, it is assumed that the operating system does not consume any processing time itself. Of course this is not true in practice, but this assumption helps in the definition that follows. In a later step, the time that the operating system has to spend on scheduling will also be taken into account.

The time t_e can be understood as the sum of all CETs in the observation period as shown in formula 6.

$$t_e = \sum_{n=1}^{N} CET_n \tag{6}$$

Here N is the number of CETs that fall within the observation period. If the observation period begins during the execution of a task, an interrupt, a thread, or a process, only those parts of the CET that fall within the observation period are taken into account and included in the total. This applies analogously to the end of the observation period. Figure 34 illustrates the CPU load calculation based on the CETs in the observation period, taking into account the truncated CETs at the beginning and end of the observation period. The CETs of tasks A, B and C are shown as gray boxes and all CETs are projected on the line "CPU". This makes it easy to see that tasks were executed for the duration of seven boxes in the observation period, which is highlighted in green and extends over twelve boxes.

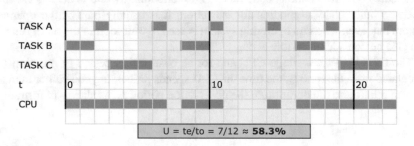

Figure 34   CPU load calculation based upon CET

When formula 4 and formula 6 are combined, formula 7 is obtained.

$$U = \frac{\sum_{n=1}^{N} CET_n}{t_o} \tag{7}$$

If you then consider a number n_C of CPUs over the same observation period, this results in the formula 8.

$$U = \frac{\sum_{n=1}^{N} CET_n}{n_C \cdot t_o} \tag{8}$$

In this case, the total load across all CPUs is the arithmetic mean of the CPU load of each individual CPU.

At the end of each observation period, the next one follows immediately. Thus, over time, one receives a large number of values for the CPU load for which—as before with the timing parameters—minimum, maximum, and average values can be collected, and histograms can be created. It makes sense to use the maximum value of all values determined for the CPU load to evaluate a system.

4.3.2   Selecting an Observation Frame

Thus far, the deliberations concerning CPU load were quite simple. The question of which observation period to choose is, however, somewhat more difficult to answer. There is, unfortunately, no clear answer to this question. The observation period depends on the project-specific configuration as mentioned earlier. In order to illustrate the effect of the choice of the observation period on the CPU load and to subsequently develop a strategy for a good choice, we will first look at a concrete example.

Using the display format introduced in Figure 34, Figure 35 shows a slightly more complex scenario. It shows four cyclical tasks with periods of 4, 8, 16, and 1000 ms. For simplicity, assume that each task has a typical CET: $CET_{4\,ms} = 1\,ms$, $CET_{8\,ms} = 2\,ms$, $CET_{16\,ms} = 3\,ms$ and $CET_{1000\,ms} = 1\,ms$. If these values are applied to formula 5, you get:

$$U = \frac{1\,ms}{4\,ms} + \frac{2\,ms}{8\,ms} + \frac{3\,ms}{16\,ms} + \frac{1\,ms}{1000\,ms} = 0.6885 = 68.85\% \qquad (9)$$

Most of the time the scheduling looks like the time interval between $t > 8\,ms$ and $t < 24\,ms$. The pattern of this time interval usually repeats itself over and over again. Then, once per second, the task with the one second period is executed resulting in the time interval as shown between $t > 0\,ms$ and $t < 16\,ms$.

Starting at $t = 35\,ms$ the CETs of the tasks deviate from their typical values and there are interruptions and, finally, at $t = 51\,ms$ the task activation of the task with a 16 ms period fails. There is obviously an overload of the system.

Now back to the question of which observation period to choose. A widely used approach is to use the superperiod of the system, i.e. the period over which the activation pattern of the tasks is repeated. In the example given, this would be $t_o = 2\,s$.

If one would calculate the CPU load on this basis with the error occurring at the position shown, the result would be $U_{2000} = 69.15\%$ (see the line labeled $t_O = 2000\,ms$ in the figure). In other words, the system is locally overloaded at one point, and yet the calculated CPU load is in a range that would be described as comfortable. In the manager view mentioned at the beginning, the traffic light would even be green despite there being a drastic runtime problem due to the failed task activation.

So the superperiod is—at least when tasks with a long period duration are involved—not a good choice for the observation period. The situation hardly gets any better if you use the period of the 'slowest' task, in this case one second, instead of the superperiod. The result can be seen in the line labeled $t_O = 1000\,ms$: $U_{1000} = 69.4\%$, also delivering a deceptive 'green' for the manager.

The subsequent two approaches with $t_O = 16\,ms$ and $t_O = 8\,ms$ are useful. Both show $U = 100\%$ for the overloaded area and provide meaningful values for the other areas.

If you select even smaller observation periods for the calculation of the CPU load, a $U = 100\%$ result still occurs for the areas with overload. However, as t_o gets

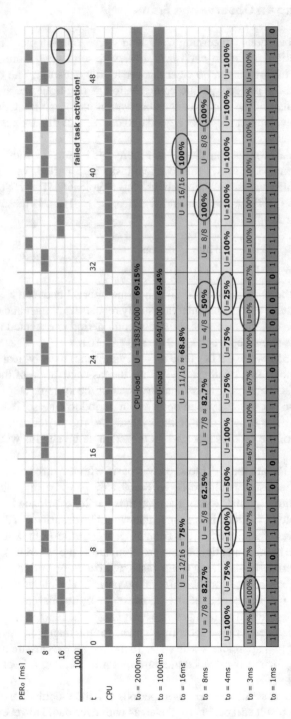

Figure 35 Impact of the selected observation period on CPU load calculation

smaller and smaller you obtain fewer usable values for the other areas, i.e. those without overload. In the extreme case of $t_o = 1$ ms the CPU load finally only jumps back and forth between 0% and 100% without providing any usable insights.

So, what generally valid strategy can we derive for the selection of a suitable observation period? In general, it should be as large as possible but not too large. As we have just shown, if the observation period is too long, the CPU load values for phases with overload will be too low. Any local overload of the system should result in the CPU load being reported as close to or equal to 100%.

For all controller-based systems, the period of the main control algorithm has proven to be a good starting point. Often this results in $t_o = 10$ ms. It is advisable to question this choice from time to time over the course of the project with regard to traces and with an understanding of the background to the CPU load calculation, adjusting it if necessary. However, any adjustments have the disadvantage that the newly calculated results cannot be compared with previously determined values.

Not all embedded systems are based on control algorithms. An airbag control unit will, best case, have undertaken only a little diagnostics and performed some network communication before it is scrapped. However, should an accident occur, a lot of code will suddenly run and must be processed in time. Diagnostic jobs during the crash are placed in a queue and can be suppressed for the duration of the crash. The crash scenarios themselves can also become quite complex. A frontal collision from an oblique angle triggers driver and passenger airbags, which may be followed by the vehicle sliding down a slope and overturning, triggering the side curtains. Which observation period for the calculation of the CPU load should be used here?

It would probably even make sense to define separate observation periods for the various scenarios. Nowhere is it written that the t_o must be the same for all operating modes and scenarios of an embedded system. This is desirable for ease of use but, if the scenarios are too different, the analysis methodology must be adapted to the state of the system.

4.3.3   Scaled CPU Load

In practice, it will usually be the case that a system no longer provides the required functionality well before a CPU load of 100% is reached. Of course, CPU load and the monitoring thereof should not be the only aspect of timing that is checked. Concrete timing requirements must also be recorded and monitored separately. The Sections 5.5 and 9.1.1 deal with this in more detail.

However, it would also be helpful if the CPU load would not indicate headroom where there is none. To use the manager's view again, a system that stops working at a CPU load of 89% places the developer in difficult position to explain the issue.

One possible solution is as simple as it is obvious. For example, if the 89% mentioned above marks the limit between 'functional' and 'non-functional', a scaled

CPU load U' could be defined that, starting from the limit U_{limit}, redefines 100%:

$$U' = \frac{U}{U_{limit}} \tag{10}$$

This is a pragmatic solution or, more precisely, a workaround. However, the underlying causes remain, unexplained. Several are conceivable here. First and foremost is the fact that important timing requirements are violated long before the 100% mark is reached and, therefore, full functionality is no longer available.

Another cause may be that the instrumentation is not sufficiently accurate if the CPU load is being calculated using CETs, and these are being measured by instrumentation. Depending on the operating system used, precise instrumentation is often not possible at all.

Ultimately, if scheduling simulation or static scheduling analysis was used, it is very likely that the model is not sufficiently accurate. This occurs when the representation of internal operating system processes and the associated runtime requirements of the operating system can only be approximately simulated or analyzed.

4.3.4 CPU Load When Using a Background Task

A background task is a task that is always executed when no other task or interrupt is being processed. If it exists, it replaces the idle function of the operating system. Its implementation can vary greatly from minimalist versions that simply increment a counter as described in Section 5.5.1.4, to versions that perform RAM checks in the background, to a background task that contains all the essential parts of the application. Anything is possible and, for each approach, there are meaningful use cases. In implementations where the background task contains all the essential parts of the application, as little as possible is undertaken in interrupts or cyclical tasks. Especially for smaller embedded systems, which focus on the implementation of state machines, this approach has proven itself because it can make very good use of the limited resources of a small processor.

When using such a background task approach, the issue with the definition of CPU load described above is that it is now permanently set at 100%. For systems that do not place any application code (functional code) in the background task at all, it can be simply considered the same as an idle loop. In other words, the computing time estimated by the background task is simply ignored when calculating the CPU load. This makes sense even if the background task contains functions like the RAM check just mentioned. Since larger sections of the RAM are usually checked, an entire run takes a considerable amount of time, so shifting it to the background task does this work in small chunks whenever nothing else is pending.

If there is a specification as to the maximum length of a complete run, this cannot be covered by the monitoring and calculation of CPU load. Instead, an alternate mechanism must be found.

Should the background task contain large sections of the actual application, all the approaches presented so far fail. A simple and practicable approach for systems of this kind could instead be implemented as follows. To get started, it is necessary to consider and then determine how long the maximum duration of an instance of the background task may be, including all interruptions. This requires use of the gross execution time (GET) and is actually one of the few sensible use cases for GET. This maximum allowed duration is then termed GET_{BGmax}. During operation, the current gross execution time GET_{BG} is now recorded with every run and the current CPU load is calculated as follows:

$$U = \frac{GET_{BG}}{GET_{BGmax}} \tag{11}$$

4.4   Bus Load

Almost everything that has been said about CPU load can also be applied to the bus load, that is the load of any communication bus. A task in a 'running' state corresponds to a message that is currently being transmitted and thus 'occupies' the bus. The overhead caused by an operating system can be likened to the information that occupies the bus in addition to the pure user data. For a CAN bus this includes the Start-Of-Frame (SOF) bit, the coding of its length in the DLC field, the checksum, etc. Formula 4 for the calculation of CPU load can be used for the calculation of bus load if t_e contains the time the bus is or was occupied during the observation period t_o.

4.5   Logical Execution Time (LET)

The *Logical Execution Time* [11] is a concept for decoupling functionality and communication with the goal of making embedded software deterministic, especially in multi-core applications, and thus more stable, secure, and easier to analyze.

The structure of a typical task follows the *IPO* model: **I**nput, **P**rocess, **O**utput. This means that at the start the task receives data, then processes this data and, before it terminates, it outputs data. The receiving and sending can also take the form of read and write accesses to memory. In particular, the time at which the data is sent depends heavily upon the execution time of the task. If it is finished earlier than usual, the data is also sent earlier. If it requires more runtime than usual, the data is sent later than usual.

With a large number of tasks on different cores of a multi-core processor, the communication between tasks quickly becomes complex, unpredictable, and sometimes unstable. Data may not always be received on time, or it may be sent twice in a period of time for which the recipient expects only a single data value. Section 6.4 discusses this case in more detail with a practical example.

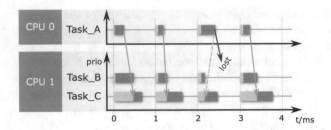

Figure 36 Communication problem due to shifts in the schedule

Figure 36 shows a communication problem between two cores of a multi-core processor. Task A on CPU 0 sends data to a Task C on CPU 1 at the end of its execution. Task C is always executed after Task B due to equal activation times and low priority.

Between $t = 2$ ms and $t = 3$ ms the execution of Task A takes a little longer than usual and that of Task B a little less. As a result, Task C will not be able to receive the expected data because Task A has not yet provided it.

With the Logical Execution Time paradigm, the receive and transmit times are decoupled to a certain extent from the execution time of the tasks. Receiving and sending take place at fixed times and in between is the execution of the task (see Figure 37).

Figure 37 The LET as the time frame in which a task may run

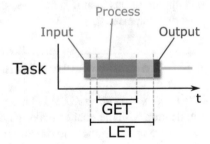

Whether this takes longer or can be processed more quickly has no influence on the defined 'pattern of communication'. In case execution takes too much time, i.e. the send time is due without the task execution having been completed, error handling occurs.

Figure 38 shows how the situation described in Figure 36 could be resolved using LET. The data provided by Task A is now always sent at the same time within each 1 ms period and received by Task C a defined time later. The latter is now no longer activated simultaneously with Task B but at the time of data reception or a short time later (Figure 38).

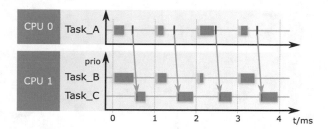

Figure 38 When using LET, communication times are decoupled from the scheduling of tasks

4.6   Summary

In the Chapter 4 various *timing parameters* were introduced. To be able to utilize them appropriately, minimum, maximum, and average values are used, as well as histograms. These were handled in the Section 4.2.

CPU load still plays an important role in practice and has therefore been taken into account in detail. It is the decisive factor in choosing the correct observation period t_o, but the optimal choice is often anything but trivial to determine. This book has tried to provide the basis for a correct decision with the Section 4.3.

With the introduction of the LET paradigm an approach was highlighted that, although it was not very widespread by mid-2020, provides clear advantages in implementing deterministic communication. This will certainly help to improve the handling of multi-core systems in particular. When, and to what extent, tool chains will universally support LET is another matter.

Timing Analysis Techniques

5

When developing embedded software, each development phase comes with its own set of timing related questions, tasks, or challenges. For example, should different operating system configurations and communication concepts be compared at a very early project phase, before the hardware is even available? Or perhaps there is a first version of the software running, but it still suffers from unresolved sporadic problems that need to be investigated? Or maybe you need to ensure that the timing remains stable and does not cause surprises during automated (regression) tests at a late stage of the project? And it may be the case that the development is finished and the timing is to be monitored by an analysis component in the regular operation of the final product.

For all these use cases there are very different timing analysis techniques available. A thorough understanding of all their possibilities, advantages, and disadvantages, as well as the necessary prerequisites for their use, is essential for efficient timing analysis. 'Efficient' here implies achieving correct timing at low cost and with as little effort invested as possible. Without the tools and approaches described here, safe embedded systems that offer high levels of availability are unthinkable.

This chapter introduces the various timing analysis techniques that are used later in the Chapter 9 in the different development phases.

The description of each timing analysis technique is rounded off by a short interview with a subject-matter expert for the respective technique.

5.1 Overview, Layered View

Figure 39 shows the timing analysis techniques that are described in detail in this chapter. The vertical axis shows the level or granularity at which the timing analysis can be performed. A detailed description is provided in the following three sections.

© The Author(s), under exclusive license to Springer Nature Switzerland AG 2021
P. Gliwa, *Embedded Software Timing*,
https://doi.org/10.1007/978-3-030-64144-3_5

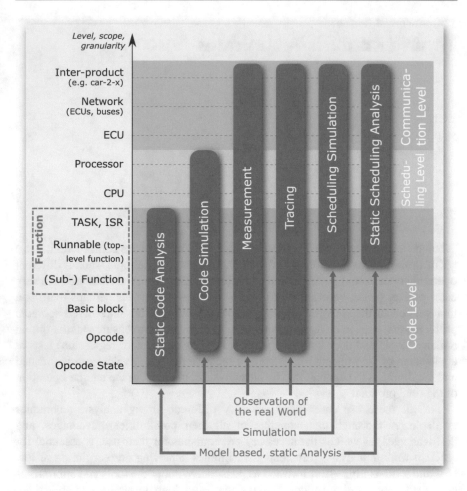

Figure 39 Overview of timing analysis techniques and their areas of application

5.1.1   Communication Level

Timing at the communication level usually concerns the elapsed time on the network's bus. Here, response times of messages, bandwidths, utilization, and buffer sizes play a significant role. The focus of attention is the central timing parameter at communication level for end-to-end elapsed time (for example, from the sensor to the actuator) or the time difference from one event in the software to another on a server.

Inter-product   Whenever the product to be developed exchanges data with the outside world, timing aspects also play a role. Take the example of Car-2-X for the networking of vehicles with each other or via the cloud to a server. One use case is where a vehicle that has detected a danger on the road warns the vehicles following it. It is obvious that this warning should not reach the receivers at some arbitrary time later.

Network   In this context, the term 'network' refers to a network of control units and the buses between them *in the product*, for example in a machine or in a vehicle.

ECU   Timing analysis in relation to a single ECU means taking a look at the processors installed in it and examining their inter-processor communication, for example via SPI. If an ECU has only a single processor, the ECU level corresponds to the processor level.

5.1.2   Scheduling Level (Also RTOS Level)

Timing at the scheduling level affects all time-related effects that relate to how the operating system organizes the software. Therefore, the scheduling level is also called the operating system or RTOS level. A key timing parameter at the scheduling level is the *response time* of tasks (see Section 4.1).

Processor   A processor is a piece of silicon that contains one or more CPUs, various memories, and peripherals (see Figure 4). The CPUs communicate with each other and often they also interfere with each other, such as when accessing shared memory. In addition, parts of the code running on different CPUs must be synchronized at certain times.
The processor level is the level at which multi-core scheduling takes place. This means that operating systems that offer multi-core scheduling (such as POSIX based operating systems) must 'keep an eye on' the entire processor and its CPUs. For example, you can migrate the processing of threads from one CPU to another.
Higher levels, on the other hand, remain outside the responsibility of the operating system. For example, it is not easily capable of starting a task on another processor or to initiate a thread on another processor.
Multi-core timing aspects are discussed in more detail in Section 7.

CPU   A CPU (Central Processing Unit) is an execution unit that can only process one thread, a chain of instructions, at a time. Hyper-threading CPUs are deliberately excluded here because they are rarely used in embedded systems. If another instruction chain is to be executed (e.g. an interrupt), the current instruction chain must be interrupted.
The CPU level is the level at which single-core scheduling, such as offered by OSEK, takes place.

5.1.3 Code Level

When analyzing the timing of elements at the code level, the focus is on their processing and the time required for this. The central timing parameter at code level is the net run time (CET, Core Execution Time).

Interruptions—for example by interrupts—are not considered at code level. In other words, if interruptions occur while considering the code level (for example, when measuring the CET), they must be deducted.

Function   The term 'function' shall be used here as a superordinate term for all function-like constructions. This includes regular C functions (for example `int giveMeANumber(void){ return 42; }`) as well as OSEK tasks or interrupt service routines.

The main reason for this extended definition is that it allows the hierarchical levels to be nicely represented, as shown from top to bottom in Figure 39.

> **TASK, ISR**   AUTOSAR CP or OSEK tasks and some of the interrupt service routines are subject to scheduling and are thus organized by the AUTOSAR CP or OSEK operating system during runtime.

> **Runnable (top-level function)**   AUTOSAR CP tasks typically call the runnables assigned to them one after the other. Like tasks, the runnables are typically `void-void` functions.
>
> However, even applications that do not use AUTOSAR operating systems often follow the concept of 'top-level functions' with which the tasks of the operating system are 'filled'.

> **(Sub-) Function**   The runnables or the functions at the top level call functions themselves, which in turn call other functions.

Basic block   All code, including functions, can be divided into basic blocks. Remember, a basic block is a series of machine instructions that are not jumped into or out of. The commands of a basic block are therefore, without exception, all processed sequentially starting with the first command.

Basic blocks already played a role in Section 2.2.

(Machine) Instruction   A machine instruction is a single instruction. It is at this level that the granularity and accuracy of the majority of tracing and measurement tools ends. An example measurement task could be: measure the CET between the execution of the instruction at address X and the execution of the instruction at address X+Y. Measurements or traces cannot be resolved any more precisely than this.

Opcode State   As the Section 2.6 has shown, each machine instruction is processed in several steps, the opcode states. In timing analysis, only static code analysis and sometimes code simulation considers effects at this level.

5.2 Definitions of Terms

Before the timing analysis techniques are explained in detail, the following is an explanation of some terms that play an important role in the subsequent sections and chapters.

5.2.1 Tracing

Tracing describes the process of recording events in a memory together with a time stamp. At a later point in time, this trace memory content can then be used to reconstruct the original events as they occurred.

Here are a few examples to illustrate this. A CAN trace allows us to analyze the communication over the CAN bus or to 'replay' it. Each message transmitted on the CAN bus is recorded together with its message ID, the user data, and a time stamp.

If an OSEK or AUTOSAR CP operating system is extended in such a way that it records the activation, start, and termination of all tasks together with the task ID and a time stamp, it is possible to later visualize and analyze when which tasks were activated, run, interrupted, etc.

Even an example far from the embedded software world, namely the logbook entries of a sailing yacht, can illustrate the concept of tracing. Among other things, every entry and exit to and from a port is logged in the logbook together with date and time. The path of the yacht can later be traced via this. The example is also helpful in that it makes it clear that the trace data is often only an excerpt of what actually happened. The path of the yacht can only be roughly reconstructed. The precise path the yacht took between the ports is not shown in the entries.

Later, we will look at scheduling tracing in more detail, i.e. the visualization of tasks and ISRs, or processes and threads, over time. In many respects, scheduling tracing is similar to an oscilloscope, and this analogy will be used several times in the further course of this book. However, instead of voltages as recorded with an oscilloscope, in scheduling tracing the states of tasks are recorded and displayed against a time axis.

Lastly, a note to avoid confusion. The terms tracing and traceability have the same root word but mean different things. Tracing was just explained. Traceability describes the ability to track different artifacts during development. For example, a test result should be associated to a test. The test itself, in turn, should be associated with the function that it is supposed to test, while that function is associated with the requirement that formed the basis for its development.

5.2.2 Profiling, Timing Measurement, and Tracing (Again)

Profiling describes the process of accessing timing parameters from a running embedded system or a simulation. Figure 40 shows two ways to do this. One approach is to perform *runtime measurements* that directly determine the desired timing parameters. The other approach is undertaken indirectly via *tracing*. When tracing, only trace data is collected initially from which the timing parameters can then be extracted.

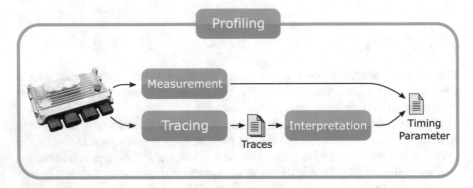

Figure 40 Two different approaches to obtain timing parameters

5.3   Static Code Analysis

'Static' here is to be understood in the sense of 'offline', i.e. it is an analysis in which the software under investigation is not being executed. There are many static code analyzers, but very few of them aim to analyze the runtime behavior or timing of the software. Among other things, there are tools for analyzing stack requirements, tools that attempt to detect software errors, and tools that check for compliance with coding guidelines.

In the following, we will limit our view to static code analysis that focuses on timing. More precisely, we will examine the determination of the maximum possible core execution time of a given code fragment, such as a function. The maximum possible CET is expressed as WCET (Worst Case Execution Time) and should actually be termed WCCET (Worst Case **Core** Execution Time). In most cases, static code analysis also determines the BCET (Best Case Execution Time), but this is less interesting because it is less critical with respect to the safety and reliability of the software.

There are basically two different types of static code analysis. One starts with the source code, the other with the executable. The former has hardly any practical use outside the academic environment, hence why only analysis based on the executable is considered in the following.

5.3.1   Fundamental Functionality and Workflow

Figure 41 shows which data is relevant when working with static code analysis. It also visualizes what the workflow looks like.

First, static code analysis reads the executable and *disassembles* it, that is, the binary machine code is translated back into the assembler instructions. See also Section 1.3.5. From the disassembled code, control flow and a function call-tree can be derived. As the name suggests, the function call tree tells you which function calls which other function(s). In addition, the analysis determines the maximum number of loop iterations.

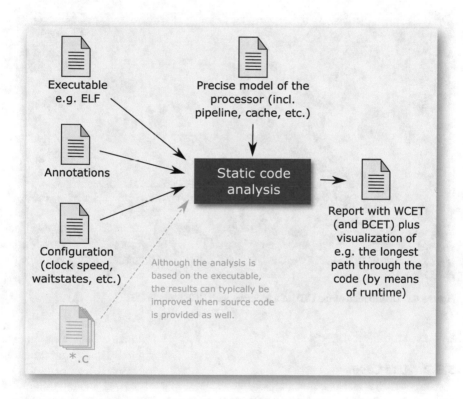

Figure 41   Workflow of static code analysis for WCET calculation

The collected data is now merged and the maximum possible runtimes are added up along the flow of control. The executable contains the memory addresses of all machine instructions and data, which is why the analysis can even consider the effect of cache and pipeline on the runtime. For this purpose, the analysis requires a very precise model of the processor in addition to the memory configuration (e.g. specification of the wait states for flash accesses). In many cases, the VHDL or

Verilog data of the processor manufacturer is used to create this model. VHDL stands for Very High Speed Integrated Circuit Hardware Description Language and can be thought of as the source code for the implementation of a processor.

If all time data are to be given in seconds or nanoseconds, the analysis also requires information on the processor clock. This results from the crystal used and the configuration of the processor's clock unit.

Finally, it should be noted that in most cases the actual WCET cannot be calculated in finite time, but it can calculate a value X which is guaranteed to be greater than the WCET. The result of the analysis is therefore always on the safe side and can be understood as a safe upper bound, see Figure 42.

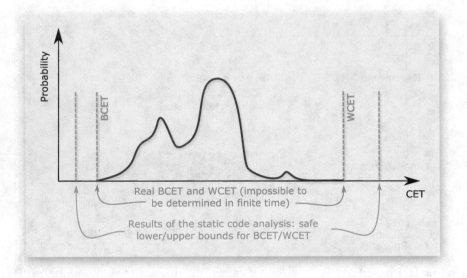

Figure 42   Distribution of the CET of a function with BCET and WCET

5.3.2   Use Cases

Static code analysis is practically unavoidable whenever it is necessary to guarantee worst case times. Some security standards regard static code analysis as 'highly recommended' (see Section 11).

Abstract Interpretation is the name given to the underlying method used in most cases for the static analysis. It is a formal method that—when applied to WCET analysis, and implemented and configured correctly—can be considered a mathematical derivation of the WCET.

It is crucial that the result of the analysis is completely independent of any test vectors used. This ensures there is no need to factor in which vectors cause the maximum runtime.

Since static code analysis uses a ready-made executable that is not executed, the analysis can be undertaken regardless of hardware availability.

Another use case is the automated WCET verification as part of the build process. Each time the software is compiled, static code analysis runs to check whether the WCETs of certain functions exceed predefined limits or whether the increase in WCETs is more than x percent over the last software release.

If the evaluation is detailed enough, the static code analysis can also be used for runtime optimization.

Figure 43 shows the results of the analysis of two functions in aiT [12], a static code analysis tool by AbsInt [13]. Proc3 is on the left and memcpy_x is on the right. In the blue boxes you can see the source code fragments. The white boxes on the paths indicate the maximum number of times the respective path is executed. If the executable contains dead code, i.e. code that cannot be executed under any circumstances, the corresponding code block is displayed in gray instead of blue and "Infeasible" appears on the corresponding paths. The path of the longest execution— the WCET path of the function—is highlighted in pink and has a blue or green arrow (the sections with green arrows appear orange with the pink background). The blue arrows mark the regular execution path and the green arrows mark the path for executed conditional jumps, such as when the condition for the jump is met. In addition, aiT provides detailed information about the time contributed by the individual code parts of the function to the WCET.

5.3.3   Limits of Static Code Analysis

Unfortunately, static code analysis is not as simple in practice as it looks. Even when building the function call tree, the analysis usually encounters problems that it cannot solve without external help, for example from the user.

5.3.3.1 Indirect Function Calls
Indirect function calls, that is function calls in which the function to be called is passed to the call as a parameter, can be difficult to analyze. In some cases it is not possible for the analysis to determine which values this parameter can take. In such a case, the function call tree is incomplete. The analysis 'knows' that further functions are to be added to it, but it remains open which functions are involved. Instead of branching to a further branch, a red question mark is placed at this point in the tree.

5.3.3.2 Recursion
The situation is similar for recursions, i.e. functions that call themselves. The interesting question here is, how deep is the recursion, i.e. how often could the function call itself at most.

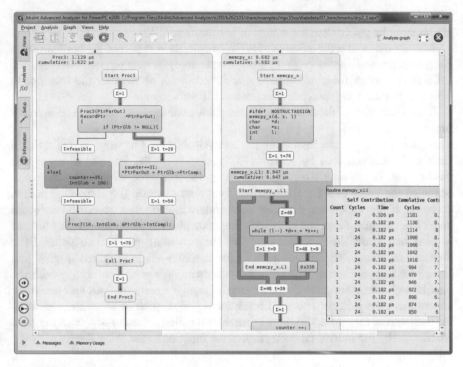

Figure 43 WCET path of two functions (tool: aiT [12] from AbsInt [13])

5.3.3.3 Upper Loop Bounds

Even if the function call tree is complete, the analysis reaches its limit for some loops and cannot determine the maximum number of loop iterations. However, this number is required to calculate the WCET. Instead of using the real upper loop bound, a value that may be far too large is used for the analysis thus leading to significant over-estimation.

5.3.3.4 Annotation

These three stumbling blocks (unresolved indirect function calls, recursions, as well as incorrectly identified upper loop bounds) must be clarified manually by the user by providing additional information. It is then said that the code must be *annotated*.

Hint The situation can be defused very easily by requiring the necessary annotations from those responsible for creating the code. For example, if a software house delivers a set of functions that contain indirect function calls, this supplier must be required to deliver annotation files in addition to the code that—similar to a specification—clearly identify e.g. which *call targets are possible for each individual indirect function call.*

Often, the call targets are only defined by a code-generating tool at a later point in the development process. In this case, the supplier of this code-generating tool must extend the generator in such a way that annotation files are generated in addition to the code.

This approach of requiring suppliers to completely annotate their deliveries was consistently introduced and successfully implemented by a large German supplier to the automotive industry several years ago.

5.3.3.5 Modes of Operation and Mutually Exclusive Code

For an application for which all indirect function calls and recursions as well as all loop bounds have been correctly recognized or annotated, additional annotation is usually still required.

Often, applications are designed to support different operating modes. In the context of an aircraft these could be 'on the ground', 'in the air', and 'control unit is being flashed'. The respective code parts are not always separated from each other in such a way that they are mutually exclusive, at least not from the perspective of static code analysis.

When examining code with a surprisingly high WCET in detail, the developer will find that there is mutually exclusive code in a path. Now they have two options. Either the code is rewritten so that the static analysis recognizes that both parts cannot be in the same path, or additional annotations must be added.

5.3.3.6 Over-Estimation

Even if the analysis can be done completely and the application modes have been annotated, the analysis result will often still be unexpectedly high. One reason for this may be that the overestimation, that is, the difference between the reported upper bound for the WCET and the actual WCET, is large.

It should also be noted that the overestimates also add up. For example, if you take an AUTOSAR CP task with 500 runnables and want to determine the WCET of this task, the analysis will choose the longest paths through all of the 500 runnables. It is then a question of probability whether or not this value is still relevant. Unfortunately, it is not possible to calculate the probability of the WCET or, more generally, the curve shown in Figure 42 for that task, occurring.

If this was possible, it would not be necessary to adapt the system based on the upper bound reported by the analysis. Instead the value for the CET corresponding to the required probability could be used.

However, this is not done in practice and therefore we must either live with these overestimates and operate the systems at a correspondingly lower level of utilization, or use a different analysis technique.

5.3.3.7 Interrupts, Multi-Core, and Transient Errors

As described, static code analysis is a code-level methodology and therefore excludes any scheduling aspects. Interrupts or access conflicts at the memory interface, that occur constantly on multi-core devices, are not considered.

An example will illustrate this. Using static code analysis, the value x is determined for a function F as the upper bound for the WCET. Assume that all prerequisites for reaching the WCET are fulfilled in an actual execution, that is, the cache and pipeline are in the worst possible state when the function is started and all data and parameters used by the function have the values required for the longest path. The WCET is reached during execution. According to the static code analysis method, this execution takes place without any interruptions or conflicts.

Now assume that this execution is interrupted by an ISR. The interrupt service routine is located in a memory area that is not present in the cache. Therefore, the code must be loaded into the cache, as must the data used by the ISR. This overwrites code and data of the function F in the cache.

After processing the ISR, the execution of function F is continued. Of course, the total CET of the ISR must be subtracted from the gross execution time (GET). The execution time of the ISR is deducted. Even if this is done, the caches are now in a 'worse' state and additional delays occur during the further execution of function F. When function F finishes, it now has an actual execution time that is *beyond* the upper limit specified by the static code analysis!

In principle, a very similar situation occurs when another CPU accesses the shared memory during the execution of function F and thus delays its execution. Considerable delays can occur in this manner, especially when accessing arrays. However, even when isolated memory areas are present, conflicts can still occur when accessing the shared internal address and data buses, such as the crossbar (Figure 4 on page 14).

For the sake of completeness, transient bugs (see Section 7.1.3 on page 192) are also ignored by static code analysis.

What do these restrictions mean in practice? A real WCET analysis with reliable results is only available for functions that are not interrupted. Interrupts, exceptions, and task preemptions must be excluded. A reliable WCET analysis for multi-core systems is practically impossible.

However, it remains a useful analysis technique in a multi-core environment (see the keyword "automated WCET verification as part of the build process" in Section 5.3.2 above).

It is also possible to move away from this strictly formal approach and to calculate the best possible upper bounds through the use of trace data [14].

5.3.4   Interview with an Expert of Static Code Analysis

The following interview is intended to round off the topic of static code analysis and to look at it from a different angle. The tools aiT and TimeWeaver are products of the company AbsInt [13], which Prof. Reinhard Wilhelm co-founded in 1998.

Peter Gliwa: Static code analysis is a broad field. At this point, we will limit ourselves to the analysis that aims to determine the BCET and WCET. Concisely summarized in one sentence, how does it work?

Prof. Reinhard Wilhelm: The WCET analyses are too complex to be described in one sentence. The initial problem is the great variability of the execution times of machine instructions. These depend on the execution state, e.g. on the cache content. For each point in a program, aiT calculates an over approximation of the set of all possible execution states. To be able to predict a cache hit on a memory block, this memory block must therefore be in all calculated cache states.

The first step of aiT is to reconstruct the control flow graph from the executable machine program. Then, aiT determines the above mentioned over approximation of the execution states. With the help of this, aiT can safely estimate the execution times of all machine commands in the program upwards.

Finally, the longest path through the program on this basis must be determined.

Peter Gliwa: What are the central use cases?

Prof. Reinhard Wilhelm: All safety critical hard real-time systems. For most of them, a determination of the WCET via time-of-flight measurements is not possible for reasons of complexity, i.e. the number of cases to be measured, and is therefore not accepted by the regulatory authorities.

aiT is often used even with less critical applications, since no test cases and test inputs have to be built, thus saving a lot of time and effort.

The technology is also used in our TimingProfiler product for code optimization at a very early stage of development when measurements on the hardware are too expensive or impossible.

Peter Gliwa: This topic has been discussed amongst academics for many years and is very well researched. How is it that it is not yet widely present in the everyday life of embedded software development?

Prof. Reinhard Wilhelm: The first publications on static run-time analysis appeared in the late 1980s but, at that time, for architectures with constant instruction execution times. In the 1990s, architectures with caches, pipelines, speculation, etc., were used for the first time whose instruction execution times were dependent on the execution state. We have solved the WCET analysis problem that arose from this and implemented a solution. The use of this technology is widespread among those users who are aware of the problem. The rest are lulled into a false sense of security and rely on methods that are fundamentally unsound.

Peter Gliwa: With a certain delay, processor features that significantly increase the average computing power are transferred from the desktop area to the embedded world. Complex cache logic or tricky branch prediction units are part of this. What impact does this have on static code analysis and what should users be aware of?

Prof. Reinhard Wilhelm: As already mentioned, this already happened in the 1990s. Most cache architectures are easy to analyze. For caches with random replacement, one cannot expect to predict such a large number of cache hits as is possible with, for example, a cache with LRU replacement. However, aiT can precisely analyze that which can be predicted for a cache with random replacement. Of course, this is not possible for measurement based methods. These would again fail due to the excessive number of cases to be considered.

However, complex architectures can often be configured by the developer to attain greater predictability. The company AbsInt offers help here.

Peter Gliwa: How are things with regard to multi-core?

Prof. Reinhard Wilhelm: WCET analysis for multi-core is more complex because, in addition to single-core analysis, you also have to do an analysis of interference on shared resources. The results of the two analyses must then be cleanly combined. aiT supports this for a range of different processors.
AbsInt also offers TimeWeaver, a hybrid tool that takes interference effects into account automatically.

Peter Gliwa: The result of the static code analysis is independent of test vectors. Thus, it can replace certain tests, which saves time. However, to get a complete call tree and to limit over estimations, annotation is required, which takes time. Is the whole thing a zero-sum game in practice?

Prof. Reinhard Wilhelm: This is what you would say if you were willing to compare apples with pears. On the one hand, you have a method which is able to provide a guarantee for punctual reaction. On the other, you have a method which at best gives you a good feeling. What would you rather have when you board a plane? The Formal Methods Supplement to the international DO178-C standard for the development of aircraft software takes this into account. The former method would be described as adequate and thus approved for certification, while the latter would not.
Moreover, aiT also saves a lot of effort and time in many cases. In order that users can determine this for their application scenario, AbsInt offers tool evaluations.

Peter Gliwa: Was there an experience in practical use that delivered an 'aha moment'?

Prof. Reinhard Wilhelm: Well, in our first analyses of the Airbus aircraft control software for the A340 and the platform flying in the A340, our results lay in the middle between the worst execution times ever observed and the results of the measurement based method previously used by Airbus, which operated with a 50% safety margin. This made the Airbus team very happy because they received a new, reliable method from us that determined more precise run-time limits than Airbus had had previously, and because we had also shown that their own run-time limits, with which the A340 had been certified, were indeed safe, although less accurate.

Peter Gliwa: Finally any recommendations or practical tips?

Prof. Reinhard Wilhelm: Always be clear whether you require a guarantee or just a good feeling, and try out our tools.

Peter Gliwa: Thank you!

Reinhard Wilhelm studied mathematics, physics, and mathematical logic at the Westfälische Wilhelms-Universität in Münster from 1965 to 1972, as well as computer science at the TU Munich and Stanford University. In 1977 he completed his doctorate in computer science at the TU Munich. Since 1978 he has been a professor of computer science at Saarland University, where he held the chair for programming languages and compiler construction until 2014. From 1990 until  2014, Wilhelm simultaneously headed the Leibniz Center for Informatics (LZI, formerly the International Meeting and Research Center for Informatics, IBFI) located in Schloss Dagstuhl as scientific director.

Wilhelm's research interests include programming languages, compiler construction, static program analysis, embedded real-time systems, and the animation and visualization of algorithms and data structures.

In 1998 he founded the spin-off AbsInt together with members of his chair in order to industrialize a correct and precise method for runtime analysis of real-time programs developed at the chair. AbsInt's tools (among others Astrée, developed under the direction of Patrick Cousot and Radhia Cousot) have been used for the certification of safety and time critical subsystems of the Airbus aircraft A380 and A350.

5.4   Code Simulation

A code simulator executes machine code for any processor on the PC (x86). Here, the PC is simulating the other processor. Since code for one platform is executed on another, the term 'cross-platform' is often used in this context. Similarly, the compiler for the target processor is also called a cross-compiler.

Compared to other timing analysis techniques, code simulation does not play such a major role, which is why this section is kept short.

Those simulators that execute the code compiled for the x86 rather than the target processor will not be considered here. Although these simulators are very interesting for function development, the results they generate regarding the runtime behavior of the software are often of little practical use.

5.4.1   Functionality and Workflow

Code simulation usually involves the examination of smaller sections of code, for example a single function or an algorithm. The simulator consists of software in which the target processor is emulated. The simulator can execute an executable generated for the target processor. The level of detail depends on the simulator used.

In the widely used *Instruction Set Simulators* that are often provided with some compilers, correct timing is typically not their main focus. Pipelines and caches are not simulated, nor are the peripherals, such as timers, MPUs, and debugging modules. Timing analysis based upon such types of simulator is correspondingly vague.

Figure 44 shows how the program in Listing 15 is compiled on the command line and then passed to the simulator—in this case, Wind River's RTASIM for the PowerPC.

The output of the simulator only reports the return value of the `main` function and the total number of executed machine instructions.

Listing 15 Minimalistic program that returns 42

```
1  int main(void)
2  {
3      return 40+2;
4  }
```

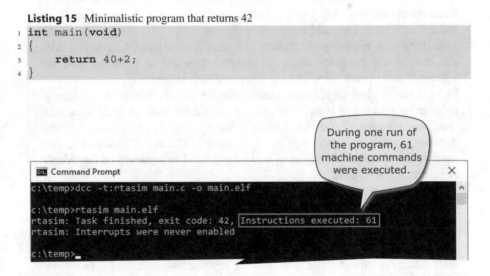

Figure 44 Translation of the program and simulation including output

5.4.2   Use Cases

In order to bring unit tests closer to the real target system without having to run the tests on real hardware, they can be compiled for the target processor and then executed using a code simulator. If this happens anyway, it is useful to store the number of instructions executed for each test. With a minimal extension of the (unit) test environment, additional rudimentary statements about the runtime requirements (or, more precisely, the CET) of the tested functions can be obtained.

If these values are recorded systematically and automatically over the course of the project, they can serve as a good indication of the net runtimes of the tested functions. If these simulated values are then compared with measured values, a simple, cost-effective runtime analysis is obtained and the net runtimes of the functions are clearly visible.

It is recommended to automatically monitor differences to the runtime of previous versions of the software during tests after changes, such as issuing a warning if it increases by x percent. If the increase is caused by additional functionality, everything is probably fine and the new reference runtime is, as a result, higher. However, if the runtime of a function jumps from one software version to the next due to a faulty 'code embellishment' measure, this is immediately noticeable. Without this kind of verification, the questionable function could find its way into a software release, be the cause of sporadic scheduling problems, and only be identified as the culprit many weeks later.

There are also much more complex and powerful simulation environments as that Synopsis provides with its virtualizer (see Figure 45). It forms part of the Virtual Prototyping Environment [15] with which an embedded system can be simulated. The example shown consists of two ECUs. The first ECU has an AURIX processor running an OSEK operating system. The second ECU distributes its functionality over two SoCs (**S**ystem **o**n **C**hip): a gateway and an Arm-based Linux environment.

In the screenshot, different views are arranged one above the other, all of which have the same horizontal time axis in common. A vertical line, such as the one shown in red, marks the same point in time in all views. The topmost view shows an OSEK task trace of CPU 0 of the AURIX processor and, below that, the function trace of the same CPU. Next is the function trace of the gateway and, below that, a trace of the SPI bus.

At the very bottom, with the black background, the console output of the multi-core embedded Linux platform is displayed.

5.4.3 Limits of Static Code Simulation

In recent years, the effort required to keep the simulation of the *complete* software executable sufficiently close to reality has been considered to far exceeded the benefit. The term 'embedded software' has its origin in the fact that the software is embedded in an environment of sensors, actuators, communication buses, and other hardware interfaces. All of this must also be simulated to a sufficient level of detail that usually involves an enormous amount of effort.

The interview at the end of this section on code simulation makes it clear that, in this respect, a lot has happened in recent years (up until 2020) and code simulation is now—or will be in the near future—used more often in the development of embedded software.

As an intermediate step to full simulation, the combination of the runtime measurement approach described above, the simple counting of instructions in connection with the unit tests, and scheduling simulation can be used. Section 5.8 on page 148 deals with scheduling simulation in detail. With this combination, fully automated as part of the build process, simple run-time validation can be performed at both the code and scheduling levels. While such validation cannot replace testing

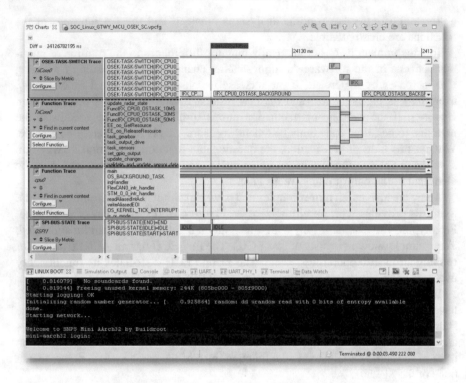

Figure 45 Synopsis' Virtualizer as part of the Virtual Prototyping Environment [15]

on real hardware, it is possible to catch different types of errors in a fully automated way before the software is flashed into the hardware.

Like all timing analysis techniques that are not based on model-based, static analysis, the results depend on the test vectors that stimulate the code to be analyzed during simulation. If the 'correct' vectors are not used here, the simulated runtimes are too low, leading to unpleasant surprises when the code is used in the real environment.

5.4.4   Interview with an Expert in the Area of Code Simulation

Like the previous section about static code analysis, this section about code simulation comes with an interview with a technical expert. The simulation environment mentioned in the interview is the Virtual Prototyping [15] environment from Synopsis [16].

Peter Gliwa: In a nutshell, how does code simulation work?

Kevin Brand: Generally speaking, there are two kinds of code simulation. One is the SIL approach, which executes 'Software In the Loop'. Most embedded software is developed using portable languages such as C, C++, or Ada that are available across platforms. In a SIL setup you compile the application software for x86 and run it in a simulation environment. There you can debug and verify it and put it into the loop of some simulator or environment that you can test it against. An example of this is the Synopsys Qtronic product. Using this approach you are, to a large extent, abstracting away the hardware dependent software and only simulating your application code.

Those types of code simulations are similar to MIL or 'Model In the Loop' simulations that you see with MATLAB in combination with Simulink. These use a high-level model to generate standards-based, AUTOSAR compliant C code for eventual compilation as application code and, hence, are also simulation-ready for a SIL simulator like Qtronic.

The other way is to take the real code, cross-compiled for the real target, and, instead of flashing it onto the target processor, you run it through an instruction set simulator. The simulator executes the code on a model of the real hardware, instruction by instruction, just like the real hardware would do. This model can be anything from a simple core/memory model to a more complete model including several processors, or even several ECUs, with all the communication between them.

Personally, I have been focusing on the instruction set simulation methodology, executing on a hardware model of the system.

Peter Gliwa: When you say 'model' you refer to the model of the processor as well as to the model of the environment including external signals, external communication, etc. Correct?

Kevin Brand: I would say yes and no. Yes, because the model can include all of that and, no, because 'external' might not be the correct word. What you refer to as external signals, external communication, and so on, does not sit outside the simulation. The core model is generally not simulating the code independently. It is sitting on the simulation back-plane or the simulation kernel if you like. In our products we use the SystemC standard, which is our backbone. The simulation kernel simulates all kinds of models at the same time, such as memory models, CPU models, peripheral models, communication models, etc.

Peter Gliwa: So, when I simulate some timer-related interrupts, these would be visible just like on the real hardware?

Kevin Brand: Yes. Any event that occurs is simulated at precise times. For instance, if your software uses a timer and a compare-match interrupt to execute code every millisecond, you would see such periodical events occur at the same intervals in time as they would on the real hardware.

Peter Gliwa: Understood. So you can simulate software not only at the code-level but also at the scheduling-level. An embedded operating system does not do much more than what you've just described. It sets up compare match interrupts at dedicated points in time, for timed activations and alarms, and executes tasks according to their states and priorities.

How about GPIOs? Can you configure the states of input pins by defining at which times they are 'high' and 'low'?

Kevin Brand: So, what you are referring to here is stimulus from the world outside the chip, the silicon, correct?

Peter Gliwa: Yes.

Kevin Brand: Again, there is no boundary on the model itself. You could have an MCU connected to an ASIC through an SPI interface. The kernel sees all of that activity, providing it is modeled accordingly.

For data that is truly coming from outside the model we have various options. Some are very simple, such as our scripting interface. With scripts you can generate signals on the kernel so that the kernel knows about them and schedules them so that they get injected into the platform and the hardware model can respond to them.

Another option is to connect the simulation model to plant models, for example, those developed in MATLAB/Simulink. Several tools, like CANoe from Vector, provide interfaces for simulations. FMI is also a common simulator interconnect methodology. We can also inject and monitor serial I/O messages like Ethernet, CAN, and SPI.

Ultimately, everything is controlled by the kernel, no matter how you hook external sources of signals or data onto the simulation. The simulation kernel is a time and event driven backbone that drives not only code simulation but the hardware interaction too

Peter Gliwa: So, when I have set up my rest-bus simulation in CANoe for my tests with the real hardware, can I use the same rest-bus simulation for the code simulation environment? Do you use virtual buses like the virtual CAN bus the Vector PC drivers come with?

Kevin Brand: Yes, you can use the same rest-bus simulation and, no, we do not use the virtual CAN bus. Instead, we interface to CANoe through shared memory and use the Vector interface APIs. The information from CANoe is then made available on a CAN, CAN-FD, Ethernet, or FlexRay bus, modeled on the simulation back-plane. The two simulations—the rest-bus simulation on the one hand and the hardware and code simulation kernel on the other—are synchronized. We call it 'V-HIL', Virtual Hardware In the Loop and you can in fact use the same rest-bus simulation for the HIL and the V-HIL.

Peter Gliwa: Yes, of course. Now that I think about it, using the Vector virtual buses would be a bad idea. If you were to set up a periodic message to be sent out every 10 ms, these would be present on the virtual bus with this period in real time. However, I guess you can run the simulation independent of real time. Correct? This brings us to the more general question of what the relationship between simulation-time and real-time looks like.

Kevin Brand: The simulator is event-driven. So the more events that happen in a given real-time frame, the longer the simulation of such a time frame will take. Basically, it comes down to that. So, the simulation speed can vary a lot dependent on the event bandwidth.

So, the variation can not only be seen between simulators, but also within the same simulation. The boot-up part of AUTOSAR, where the software is initializing the peripherals and everything is configured, generally generates a lot of events. Once it gets to normal execution the simulation typically runs a lot faster.

As a result of this variation in speed, it is often difficult for us to connect to real-time interfaces that demand a certain throughput to be maintained.

Peter Gliwa: Did I get this right? Depending on the complexity of what is simulated, the simulation runs either faster or slower than real time? If I have a rather simple system, can I slow down the simulation to run in real time?

Kevin Brand: We do actually have a block you can insert which then throttles down simulation speed. This is done when you want to interface to real-time I/O outside the simulation and you are running faster than real time. Examples include USB or Ethernet.

The simulation speed can easily reach multiple times real time because, in some modes, the cores go into a sleep mode and then you see huge 'jumps' in time because only very few events occur. In sleep-mode you typically still see peripheral timers increment now and then, for example, and that helps with the throttling of simulation speed. But, generally, if core models are not actively executing instructions during a simulation, the simulation event processing can be extremely fast.

Peter Gliwa: I assume whenever the software has some sort of graphical interface, for example an RPM gauge, you would want the simulation to run close to real time to avoid the indicator making unrealistic movements, correct?

Kevin Brand: Yes, so for those cases real time or slower is more desirable.

Peter Gliwa: So, with respect to what happens when the simulation can be accurate, how about the accuracy of the model of the CPU itself?

Kevin Brand: It depends on which of our platforms you use. Not all of them have a high accuracy. Some offer rather simple instruction set simulation only, but the platforms offering what we call 'fast time technology' provide a very detailed level. They simulate arbitration on processor internal buses, pre-fetch buffering, pipelines (including branch prediction), caches with complex eviction schemes, and so forth.

Peter Gliwa: What are the use cases for code simulation? Do you see it mostly being used at the code-level or the scheduling-level? Or both?

Kevin Brand: Not so much at the scheduling-level. The focus is really more at the code-level, such as understanding and verifying functional software for example. Let me give you one example, one use case that covers both levels to some extent. Today's ECUs have thousands of symbols: variables and functions. However, the hardware offers only limited local fast memories. With code simulation you can analyze which symbol gets accessed from where and how frequently. Let's say variable a resides in the core-local RAM of core 1, but rarely gets accessed by the code running on this core. At the same time, code running on core 0 reads the variable 10,000 times per second. Since the cross-core read accesses come with greater execution time, it is obvious you should have allocated variable a in the core-local memory of core 0 rather than core 1.

Peter Gliwa: With this great level of detail, is Worst Case Execution Time, WCET, a topic that customers address using your tools?

Kevin Brand: Not specifically worst case. What we have seen is the determination of headroom through code simulation. For such set-ups we have been asked to add a fake scenario on top of the existing code. We artificially increase the load without modifying the software itself. Executing the software with the additional load lets you then check if it is still safely fully functional.

Peter Gliwa: Do you have any final recommendations, advice, or tips regarding code simulation in the context of timing analysis?

Kevin Brand: Code simulation in conjunction with existing tools supplementing your hardware gives you the visibility on your system that you need for efficient debugging and testing. Your system becomes a white box.

Peter Gliwa: Thank you very much!

Kevin Brand obtained a Masters Degree in Digital Hardware Design at Edinburgh Napier University and initially worked in industry as a Hardware and System Designer. Kevin is a Senior Manager of Applications Engineering at Synopsys. He has been in the field of virtual prototyping for 15 years. With a strong focus on automotive, he has been working in research and development and, in more recent years, directly enabling and deploying virtual prototyping solutions at semiconductor, Tier-1, and OEM companies worldwide.

Synopsys provides advanced technologies for chip design, verification, IP integration, and software security and quality testing.

5.5 Timing Measurement

Runtime measurement by toggling a port pin is probably the oldest timing analysis technique. However, this does not mean that you can equate 'runtime measurement' with 'pin toggling'. There are highly optimized and accurate tools available that do not require port pins or additional hardware. If they also meet high safety requirements, they can be used to monitor the runtime in the final product during operation. Thus, they are also used as part of the system's safety concept.

From simple pin toggling to certified runtime measurement and runtime monitoring techniques, there are a range of possible runtime measurements, and there are a number of approaches that lie somewhere between these two extremes.

5.5.1 Basic Functionality and Workflow

First, let's look at the code and scheduling level. The runtimes to be measured thus refer to code (tasks, ISRs, functions, etc.) and not to signals in networks.

The runtime measurement as shown in Figure 40 is almost always a measurement based upon software instrumentation of the code. Hardware-based tracers, sometimes inaccurately called debuggers, also provide timing parameters, but they do so using previously recorded traces (refer again to Figure 40). Hardware-based tracing is described in the following Section 5.6.

The original and simplest form of runtime measurement using software instrumentation involves providing additional code at the beginning and end of the element to be measured, for example, of a function or a loop. At the beginning, the code will set a port pin to logical one and at the end it will set it back to logical zero. If the signal at the port pin is now visualized and measured by means of an oscilloscope or logic analyzer, the time the signal remains at logic one represents the gross execution time (GET) of the measured element.

Listing 16 shows a small example program that calls the root function sqrt for all positive integer values of the data type **unsigned short**. Before each call, the runtime measurement is started with StartObservation and, after returning, the measurement is terminated with StopObservation. At the start of the main function, the measurement is initialized by calling InitObservation.

Listing 16 Program that measures the execution time of library function sqrt

```
1  #include <math.h>
2  #include "Observation.h"
3
4  unsigned short result;
5
6  int main(void)
7  {
8      unsigned short i = 0;
9      InitObservation();
10     do {
```

```
11          StartObservation();
12          result = (unsigned short) sqrt(i);
13          StopObservation();
14          i++;
15     } while (i != 0);
16     return 0;
17 }
```

The actual implementation of the pin toggle can be found in the header observation.h, to be found in Listing 17. In this case the code is implemented for the Microchip AVR processor and is very simple.

Listing 17 observation.h: Implementation of pin toggle instrumentation

```
1  #ifndef OBSERVATION_H_
2  #define OBSERVATION_H_
3
4  #include <avr/io.h>
5
6  inline void InitObservation(void)
7  {
8      DDRB = (1<<PB0); // configure PB0 as output pin
9  }
10
11 inline void StartObservation(void)
12 {
13     PORTB |= (1<<PB0); // set pin high
14 }
15
16 inline void StopObservation(void)
17 {
18     PORTB &= ~(1<<PB0); // set pin low
19 }
20
21 #endif /* OBSERVATION_H_ */
```

The setup of the measurement is shown in Figure 46. Since the frequencies are typically in the range of a few kilohertz to a few megahertz, a very simple oscilloscope or logic analyzer is more than sufficient. Figure 47 shows the voltage curve for the port pin.

Very little time passes between the individual calls of the measured function, which is why the signal only remains in the low state for a very short amount of time. The duration that the port pin is in the high state represents the gross runtime—i.e. the GET—of sqrt plus half the measurement overhead.

So, why was "plus half the measuring overhead" stated? Let us assume that the two functions StartObservation and StopObservation require the same amount of runtime and the level change at the pin occurs with the same delay relative to their call, namely after the time t_1. The time which elapses after the level change until the end of the respective measurement function is the time period t_2. Each of the two functions therefore estimates an overhead of $CET_{OH} = t_1 + t_2$. The resultant duration, to which the high level at the pin corresponds, is made up of the end of StartObservation (t_2), the GET of sqrt, and the beginning of StopObservation

Figure 46 Measurement setup for the GET measurement using a pin toggle

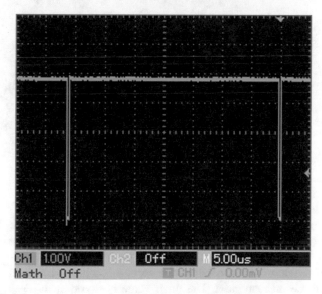

Figure 47 Signal when measuring the GET of the example function

(t_1). This corresponds to the runtime of *one* of the instrumentation functions and thus half of the overhead.

Figure 48 illustrates the measurement of half overhead. All that is required is to modify the original code of the `main` function so that no code is executed between the measurement functions. It is sufficient to temporarily comment out or remove line 12 "`result = (unsigned short) sqrt(i);`". Now the port pin is set and immediately reset, which takes 125 ns in this case. With the ATmega128 and 16 MHz crystal used here, this corresponds to two processor cycles. During measurements this half overhead, i.e. the 125 ns, can be subtracted from the result to get a very accurate measurement.

Because the implementation prefixes the functions with `inline`, the compiler does not generate any function calls at all but places exactly one machine instruction in the code. This explains why the overhead is so low.

Further down, in the Section 8.3.2 of the Chapter 8, the `sqrt` function will again play a role as we optimize this function for a reduced runtime.

Figure 48 Signal when identifying measurement overhead

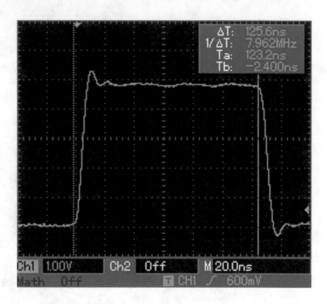

5.5.1.1 Measuring Without Port-Pins

Port pins or external measurement approaches are not always available. In addition, the handling of measuring devices is cumbersome and cannot always be comfortably automated.

Instead of the port pin and oscilloscope approach, a hardware timer can be used as a basis for the measurement. The timer value is stored at the beginning of the element to be measured, read out again at the end, and the difference between the two values reflects the gross run time in timer ticks. Over the duration of a tick—

the inverse of the frequency of the timer—the conversion into seconds can then be performed.

Based on the example already measured using pin toggling, the instrumentation will be adjusted in the following example so that the runtime of the function sqrt is determined exclusively by software. Listing 18 shows the new implementation of the header observation.h and Listing 19 shows the new C module observation.c that is now required. The code of the main function remains untouched, so it is identical to the code from Listing 16.

Listing 18 observation.h: Runtime measurement based on software only

```
1  #ifndef OBSERVATION_H_
2  #define OBSERVATION_H_
3
4  #include <avr/io.h>
5  #include <avr/interrupt.h>
6
7  extern unsigned short startTime;
8  extern unsigned short grossExecutionTime;
9
10 void InitObservation(void);
11
12 inline unsigned short SafeReadTCNT1(void)
13 {
14     unsigned char sreg;
15     unsigned short retVal;
16
17     sreg = SREG; // save interrupt lock status
18     cli(); // disable all interrupts
19     retVal = TCNT1;
20     SREG = sreg; // restore interrupt lock status
21     return retVal;
22 }
23
24 inline void StartObservation(void)
25 {
26     startTime = SafeReadTCNT1();
27 }
28
29 inline void StopObservation(void)
30 {
31     grossExecutionTime = SafeReadTCNT1() - startTime;
32 }
33
34 inline unsigned short GetGrossExecutionTime(void)
35 {
36     return grossExecutionTime;
37 }
38
39 #endif /* OBSERVATION_H_ */
```

Listing 19 `observation.c`: Runtime measurement based on software only

```
1  #include "Observation.h"
2
3  unsigned short startTime; // of the measurement
4  unsigned short grossExecutionTime; // GET
5
6  void InitObservation(void)
7  {
8      // !!! careful !!! function relies on reset values
9      TCCR1B = (1 << CS10); // timer start running at full speed
10 }
```

The result of the last measurement can be retrieved using the function `GetGrossExecutionTime`. Only the conversion of timerticks into seconds is missing:

$$t_{sec} = \frac{t_{ticks} \cdot Prescaler}{f_{SYS}} \tag{12}$$

The system clock f_{SYS} can correspond to the oscillator frequency on simple processors; on more complex processors a PLL will be used to feed the system clock. The system clock is then slowed down again by the factor $Prescaler$. Please note that there can be several prescalers (dividers) that have to be considered.

For both measurements, i.e. the one using a pin toggle and the one using a timer, an ATmega128 with 16 MHz crystal was used ($f_{SYS} = 16$ MHz). The function `InitObservation` from Listing 19 starts the 16-bit wide timer 1 without a divider ($Prescaler = 1$). As already shown by determining half the overhead with the pin toggle approach, a similar measurement error can be determined for this software based measurement. This will be subtracted from the measurements to get a more accurate result. For this purpose, an 'empty measurement' is carried out again, i.e. `StartObservation` and `StopObservation` are called directly one after the other. For the environment mentioned with the example shown, one receives a measurement error of $t_{OH} = 9$ ticks.

Since interrupt sources were disabled during the measurements, the gross runtime GET corresponds to the net runtime CET and the following section will go into this aspect in more detail.

The actual measurement provides uncorrected values between 123 ticks and 655 ticks or corrected 114 ticks and 646 ticks for the net runtime of `sqrt`. This corresponds to a gross runtime between 7.125 and 40.375 µs. By the way, no hardware is needed for this measurement. The freely available Atmel Studio 7 offers, in addition to an editor, project management, a compiler and a simulator that also models the timers correctly [17].

The minimum and maximum values can be determined by adding separate variables that are updated after each measurement if necessary. These are much like low and high water marks that indicate low and high tide levels. See `GETmin` and `GETmax` in the correspondingly adapted file `main.c` in Listing 20.

Listing 20 main.c providing minimum, maximum and average values

```
1  #include <math.h>
2  #include <limits.h>
3  #include "Observation.h"
4
5  volatile unsigned short result, iGETmax, iGETmin;
6  unsigned short GETmax = 0;
7  unsigned short GETmin = UINT_MAX;
8  unsigned long GETavg = 0;
9
10 int main(void)
11 {
12     unsigned short i = 0;
13     InitObservation();
14     do {
15         StartObservation();
16         result = (unsigned short) sqrt(i);
17         StopObservation();
18         GETavg += GetGrossExecutionTime();
19         if(GetGrossExecutionTime() > GETmax) {
20             GETmax = GetGrossExecutionTime();
21             iGETmax = i;
22         }
23         if(GetGrossExecutionTime() < GETmin) {
24             GETmin = GetGrossExecutionTime();
25             iGETmin = i;
26         }
27         i++;
28     } while (i != 0);
29     GETavg >>= 16;
30     return 0;
31 }
```

As with the pin toggle, the software-based runtime measurement functions StartObservation and StopObservation were implemented as inline functions to minimize overhead. This is at the expense of a clean software structure since the interface, i.e. the file observation.h, is more than just an interface. It also contains the implementation and reveals internal variables, such as the variable startTime.

If positive integers and corresponding types such as **unsigned short** are used in all calculations based upon (timer) ticks, there is no need to worry about timer overflows. Even if an overflow occurs between the start and end of the measurement, the difference will provide the desired value. The following example should clarify this briefly.

Assume that a timer with 16 bits is used for the measurement—just like the timer 1 of the ATmega128. Alternatively, you can use any contiguous 16 bits of a wider timer as, usually, there is a very efficient way to read just the lower 16 bits of a 32-bit timer. The upper 16 bits are simply ignored in this case.

First, consider a situation where there is no overflow between the instrumentation points. The timer value at the start of measurement $T_B = 0x1A0C$ and at end of measurement $T_E = 0x1D2E$. The difference is $T_E - T_B = 0x0322 = 802$. The corresponding source code might resemble the following:

```
unsigned short grossExecutionTime = T_E - T_B;
```

If the timer is configured to count up by one every microsecond, a gross run time of 802 μs has been measured. In a second measurement T_B has the value $0xFE03$ and T_E has the value $0x0125$. Since $T_E < T_B$ there has obviously been an overflow between the start and end of the measurement. If an unsigned 16-bit type is used as the data type throughout, the difference can still be formed using the same calculation as before. For the values from the example you again get $T_E - T_B = 0x0322 = 802$. The only restriction that exists is that a *maximum of one* overflow is allowed between the start and end of the measurement. If this cannot be guaranteed, i.e. the time span to be measured is too long, the timer must be configured to count up more slowly. Alternatively, another 16-bit wide section of a wider timer could be used. The maximum possible measurement time is the time span T_{OV} between two overflows. For a timer of width B with a tick duration of T_{tick} it is calculated as follows:

$$T_{OV} = 2^B \cdot T_{tick} \tag{13}$$

Of course, the whole thing only works if the selected data type actually provides the desired width. For most architectures `unsigned short` returns an unsigned 16-bit value. To be on the safe side, the C99 data type `uint16_t` can be used, provided that at least version C99 of the programming language is used.

5.5.1.2 Measuring the Net Runtime CET

If only the element to be measured is instrumented and measured (such as a function), it is not apparent from the signal (pin toggle) or the time difference (timer) whether one or more interruptions by tasks of higher priority, interrupts or exceptions have occurred. If there is a possibility of an interruption, the CET cannot easily be derived.

Nevertheless, in order to obtain net runtimes, interrupts can be disabled for the duration of the measurement. Exceptions would not be affected by the block, but their execution represents a special exceptional situation of the software and usually the most reasonable solution would be to simply discard all measurement data in the case of exceptions.

If the net runtime of the element to be measured is somewhat longer, the duration of the interrupt disable can quickly become so long that the resultant delay of interrupts causes the software to no longer function correctly. Most operating systems use interrupts for scheduling and, thus, preemptive tasks are also affected by a global interrupt lock.

Another approach to get net runtimes by measurement is to instrument all tasks and ISRs that could interrupt. In most cases it will be useful to simply instrument *all* tasks and interrupts. During or after a measurement, all interrupts can now be reconstructed and deducted. This leaves the net runtime you are looking for.

5.5.1.3 OSEK PreTaskHook/PostTaskHook

OSEK introduced a mechanism for runtime measurement that is also available for AUTOSAR CP and is used by various operating system vendors. This is the PreTaskHook/PostTaskHook mechanism. Before the application code of a task is called, the PreTaskHook runs and, at the end of the task, the PostTaskHook. So far, this is intuitive, because this is exactly what you would expect judging by the name of the hooks. However, the hooks are also called by the operating system during interruptions, as shown in Figure 49. Actually, the hooks should be called PreExecutionHook and PostExecutionHook.

Figure 49 The OSEK Pre- and PostTaskHooks

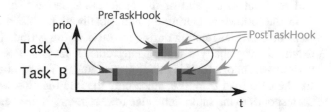

These OSEK Pre- and PostTaskHooks are only partially suitable for runtime measurement. Although they can be used to quickly implement instrumentation that also determines core execution times, the following reasons are arguments against doing so.

For one thing, the hooks have no parameters, and therefore do not 'know' which task they were initiated by. It is difficult to understand why OSEK specified this at the time. After all, the operating system knows the task in question at the time the hooks are called and it would be easy to pass this information on to the hooks via parameters. So, with the standard as it is, the hooks need to first determine which task is affected by calling the function GetTaskId(...) and this costs valuable runtime.

Another reason for not using the OSEK Pre- and PostTaskHooks is that the hooks are called frequently. If a task is interrupted ten times by another task, this results in 42 calls of the hooks just for these two tasks.

The OSEK standard itself provides the final reason. It strongly recommends for safety reasons that the hooks should only be used during development and not in the finished product.

5.5.1.4 Measuring the CPU Load Using an Idle Loop Counter

The Section 4.3 explained how to calculate the CPU load. There are basically two ways to measure it: within an observation period t_o either the idle time t_{idle} is recorded, or all the net runtimes are summed together in t_e. Formula 4 can be extended with regard to the measurement of the idle time as follows:

$$U = \frac{t_e}{t_o} = \frac{1 - t_{idle}}{t_o} \tag{14}$$

A fairly common approach to measuring idle time is to use a counter in the idle loop, as shown in Listing 21.

Listing 21 Loop body of the idle function for CPU load measurement

```
1  __isync(); /* Reset pipeline and cache */
2  if( 0 != (idleCounter + 1) ) /* saturation: avoid overflow */
3  {
4      idleCounter++;
5  }
```

Here the counter value is set to zero at the beginning of the observation period and read out at the end of the observation period. This value is referred to as Z in the following. During an initial calibration, this is undertaken without any tasks or interrupts running during the observation period. This can be achieved, for example, by globally disabling interrupts. It may also be necessary to deactivate various watchdogs for this initial measurement. The resultant value Z_0 is the value for the CPU load $U = 0\%$, since only the idle function was executed. If the value corresponds to the saturation value (0xFFFFFFFF if **unsigned int** was selected as the data type for the counter on a 32-bit processor), the observation period is too large. It should either be reduced or delay elements should be added to the loop body, such as more __isync statements.

Now the global interrupt lock can be removed again and, if necessary, any watchdogs are rearmed. From now on, after acquiring the counter value Z at the end of the observation period, the CPU load can be calculated as follows:

$$U = 1 - \frac{Z}{Z_0} \tag{15}$$

A value of $Z = 0$ indicates that the idle loop has not been run once, which corresponds to a CPU load of 100%. If $Z > Z_0$, recalibration is required. This can happen if, due to memory reorganization, the idle loop is suddenly executed from a faster memory than the original calibration.

Finally, a few words regarding the __isync statement. This is specific to Infineon TriCore architectures such as AURIX. Similar instructions exist for other architectures as well. This statement resets the pipeline and cache, which means that the subsequent statements—incrementing the counter—always take the same amount of time, regardless of the code that was executed previously. This is important when measuring CPU load using idle loop counters. Without this statement, the handful of subsequent instructions would be processed very quickly during calibration as the pipeline, cache, and branch prediction unit are almost ideally suited to execute this code very quickly. After the calibration, and while the system is running, the idle loop is continuously interrupted and, when it returns, the pipeline, cache, and branch prediction unit are probably not in a state that allows the idle loop to be executed as quickly as possible. The result would be a highly non-linear behavior of formula 15.

5.5.1.5 Measurements Using 'Performance Counters'

Many processors provide hardware features enabling determination of parameters during code execution that are closely related to timing. These include the number of cache misses and the number of conflicts during (simultaneous) access of multiple cores to shared memory, pipeline stalls, and so on.

The configuration and evaluation can be done either using self-developed software modules, by the debugger, or by tools especially developed for this purpose. The latter includes, for example, the "Embedded Profiler" [18] from TASKING for the Infineon AURIX.

5.5.1.6 Measurement Using 'Ping'

The descriptions of runtime measurement techniques are rounded off by a reference to the network diagnostic tool *ping*. It measures the response time of a simple request (more precisely, an 'echo request') from a client across a network to a host and back. This explains why the runtime measurement in Figure 39 extends to the 'inter-product' level. Figure 50 shows its usage.

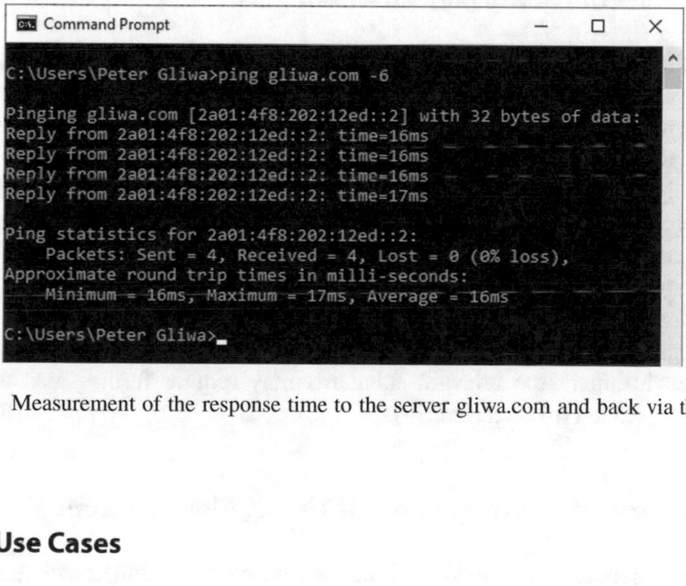

```
C:\Users\Peter Gliwa>ping gliwa.com -6

Pinging gliwa.com [2a01:4f8:202:12ed::2] with 32 bytes of data:
Reply from 2a01:4f8:202:12ed::2: time=16ms
Reply from 2a01:4f8:202:12ed::2: time=16ms
Reply from 2a01:4f8:202:12ed::2: time=16ms
Reply from 2a01:4f8:202:12ed::2: time=17ms

Ping statistics for 2a01:4f8:202:12ed::2:
    Packets: Sent = 4, Received = 4, Lost = 0 (0% loss),
Approximate round trip times in milli-seconds:
    Minimum = 16ms, Maximum = 17ms, Average = 16ms

C:\Users\Peter Gliwa>
```

Figure 50 Measurement of the response time to the server gliwa.com and back via the Internet using *ping*

5.5.2   Use Cases

There are various situations where time-of-flight measurement is the optimal solution.

The approach described in the Section 5.4.2 to determine runtimes of the tested functions in parallel to the unit tests can also be implemented by means of runtime measurements. So-called PIL setups (PIL = **P**rocessor **I**n the **L**oop) allow unit tests and runtime measurements on the real target processor with a corresponding increase in the meaningfulness of the results.

Because it is implemented quite quickly and simply, it is always the tool of choice when timing parameters need to be determined in a simple and cost-effective manner. This is especially the case when considering small and non-safety relevant embedded systems.

However, even in complex and safety-relevant projects, one is well advised to take a look at the *actual* running times occurring during operation. Whether this is performed by runtime measurement or, even better, by tracing, a look at the real system to check the timing is essential. Section 6.6 shows what can happen if the view of reality is missing and Section 9.6 provides the theoretical background to the topic.

Last but not least, we would like to highlight out, once again, that runtime measurement should not be considered solely as a pure development tool but as a part of the embedded software that is also used in the final version of the software (in the product). In this context it not only performs runtime measurement; it can function as a *run-time monitor*, comparing measurement results to ensure they lie within previously defined limits. Should a limit be violated, the embedded system can react accordingly, such as by switching to a safe state (fail-safe).

5.5.3   Limits of Timing Measurement

If the runtime measurement is to be used for profiling, it must always be remembered that the results depend on the test vectors. If the goal is to determine the maximum CET, this will only occur if all relevant parameters, variables, etc., are selected in such a way that the longest path is traversed. And even then, the pipeline and cache are certainly not in a state that would produce the theoretically possible WCET.

This problem can be largely mitigated by measuring *continuously* and storing the maximum runtimes in non-volatile memories. The runtime measurement here is thus moving from a development tool to being an integral part of the software that is also active in the final product.

Although this approach leads to highly reliable values in practice, in a safety-relevant environment the relevant standards may require further measures to be taken (see Section 11.4 on page 295).

5.5.4   Interview with an Expert of Timing Measurement

The previous sections on analysis techniques were each concluded with an interview with an expert on the technique concerned. This section, "Timing measurement", is combined with the interview covering the topic "Instrumentation-based tracing" in Section 5.7.4.

5.6 Hardware-Based Tracing

As a way of approaching this topic, we start by a look at the past. A long time ago, processors had external memories for the program code and often also for the data. These external memory devices were coupled to the processor via address and data buses together with their control lines. To visualize the program flow, a logic analyzer could now be connected to the address and data buses as well as the control lines. Thus, all memory accesses became visible. In powerful logic analyzers the functionality of the control lines and, in some of them, even the opcodes were stored so that the executed assembler commands together with the exchanged data became visible on the screen.

With the introduction of internal memories this approach was no longer directly applicable, since the necessary signals could no longer be accessed outside the chip. The era of emulators began with special versions of the processors in which the desired signals to be monitored were routed to the outside. These emulator chips naturally had considerably more pins than the normal version of the processor and were usually mounted on special adapter boards, exposing the same pins at the bottom in the same arrangement as the normal version. This allowed these 'emulator heads' to be placed on the target board and made it possible to observe the software at work in its intended environment. The logic analyzers were replaced by the emulators, which can be seen as a combination of emulator chip and logic analyzer including a disassembler adapted to the processor used.

As clock frequencies increased, the additional lines that an emulator setup required became a challenge. The background of this electromagnetic compatibility (EMC) problem is described in detail in Section 7.1 on page 189.

The solution was to transfer the actual trace logic from the emulator to the chip. A chip with 'on-chip debug support' or 'on chip trace support' is a little more expensive because the additional functionality of course results in an increase in silicon area. Despite this, the approach is now simply accepted. Some chip manufacturers also offer differently equipped but pin-compatible versions of their processors. For example, Infineon, with their first generation AURIX, offer both regular 'production devices' as well as 'emulation devices' that feature, among other things, the on-chip trace logic and large trace memory.

Even though an essential part of the debug and trace logic is on the processor, this does not mean that you do not need additional hardware for debugging and tracing. A debug and trace tool is placed between the processor and the PC and implements the protocol of the debug and trace logic on the chip, transferring data in both directions. Very-high bandwidths often occur, especially during tracing, so that the EMC problem is not completely eliminated, even with this approach.

5.6.1   Basic Functionality and Workflow

The mode of operation is described according to the type of tracing undertaken, which can be either tracing, focusing on the program flow, or the tracing of data.

5.6.1.1 Instruction Tracing for Program-Flow Analysis

In the following, we will examine how hardware-based tracing functions. We start with instruction or flow trace, where the goal is to trace every command or instruction that has been executed. The basic setup of hardware-based tracing is shown in the upper third of Figure 55 on page 132.

Figure 51 shows a code fragment taken out of context. Framed in red are the basic blocks, each ending with a jump or function call. Any disruption of the sequential execution of instructions, other than handling interrupts or exceptions, is called a branch. Thus all kinds of jumps and calls—conditional or unconditional—are branches.

Suppose the code is being executed and is being 'watched' by a hardware-based trace tool. The blue arrows indicate the commands that are being executed. Obviously, register D15 has the value 2 because the command `jeq d15,#2,.L10` (jump to label .L10 if d15 equals 2) actually triggers a jump. The commands at jump label .L10 are executed in sequence, with the last one a call function to the function `CAN_PreWrite`.

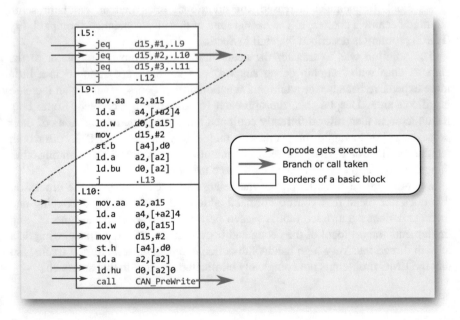

Figure 51   Code with basic blocks and jumps/function calls in red

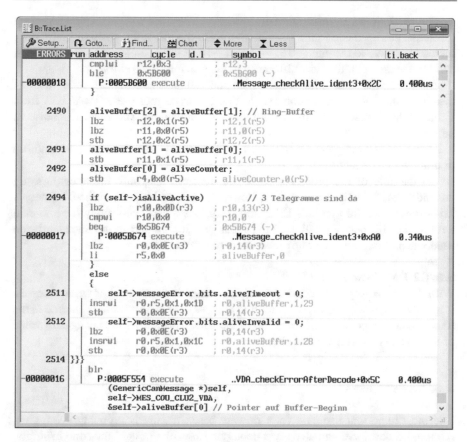

Figure 52   Screen-shot of an instruction trace (also referred to as 'flow trace')

If the trace unit on the processor were now to make an entry in the trace memory (including timestamp) for every command executed (blue arrows), the resulting trace bandwidth would be extremely high. This is not necessary to reconstruct the program flow. It is sufficient to just trace the *conditional branches taken*. All instructions in-between can be interpolated and their execution time, the timestamp, can be approximated.

Precisely how exactly the trace logic works depends on the processor used. The implementations differ in detail, but what they all have in common is that only a few decisive events are recorded in the trace memory and the majority of the executed instructions are interpolated, that is, reconstructed.

The result is an image of the observed program flow, a list of all processed machine instructions, and the timestamp for each instruction. If symbol information and source files are available, the program flow can be traced in the source code. The user can mentally 'walk' through the source code step by step and determine the program flow.

Figure 52 shows an instruction or flow trace. The lines with black text represent the actual entries made in the trace memory. On the left is the position in the trace memory relative to the end (-16, -17 and -18 in the figure), the opcode address, and, on the far right, the time difference to the previous trace entry.

All lines with blue line numbers on the left side are source code lines of the program. The machine commands that the trace tool (here TRACE32 from Lauterbach) disassembled from the memory contents are shown in turquoise in between.

Depending on the processor and trace tool used, anything from a few seconds up to minutes can be recorded.

On the side of the processor there are different hardware interfaces, mostly dependent on the chip manufacturer. PowerPCs offer a Nexus interface, Infineon's TriCore architecture provides DAP, and so on. As a user, you rarely have to deal with the details, but you will need to exchange at least a part of the trace hardware used when switching from one architecture to another.

5.6.1.2 Data Trace

With data trace, a specific address or address range is monitored for write accesses. If a write access is made to the monitored address or area, this event is entered in the trace memory together with the value of the data. This allows the changes in value of the data to be reconstructed over time.

5.6.2 Use Cases

The most important use case for hardware based tracing is probably debugging. Although a developer can also use a debugger to 'walk through' the code in single-step mode, the processor must be stopped at each step for this purpose. However, this approach this is usually of limited use, especially in embedded systems that are embedded in an environment of sensors, actuators, buses, and processors with all their real-time requirements.

Hardware-based tracing provides a remedy: it can observe the software to be examined during operation—i.e. without stopping it—and log the execution path. Unlike software-based tracing, no modification of the software is required to achieve this.

In addition to the 'debugging' use case, hardware-based tracing is ideal for runtime analysis. All the timing parameters for the functions that were executed during tracing can be determined. Usually, there is a view that displays all executed functions sorted by their CPU load, which is very useful for runtime optimization (see also Section 5.2.2 on page 87).

How do things look at the levels above the functions?

As part of the collection of OSEK specifications there is ORTI [19], the "OSEK Run-Time Interface", which is discussed in more detail in Section 10.4. Summarized in a single sentence, ORTI brings 'OS awareness' to debuggers and hardware tracers. Using ORTI, these tools 'know' that the software to be analyzed uses an operating system, and how to determine the currently running task. In most cases, the internal operating system variable which identifies the task currently running is simply recorded by means of data trace.

Hardware tracers equipped in this way can be used for the analysis of scheduling and can display runtime situations at task level, or determine the CETs of tasks. The lower part of Figure 53 shows such a representation that, in this case, even includes the runnables of the application. The upper part shows the result of a trace-based code coverage analysis.

A detailed code coverage analysis is also shown in Figure 54. The source code is shown in the large window in blue, while the machine code is shown in black on the right-hand side. To the left of it are the line numbers for the source code and the program addresses for the machine instructions. The column 'Coverage' indicates whether the respective instructions were executed during the trace as well as whether conditional jumps were executed or not.

This excerpt is originally an example to illustrate the difference between Statement Coverage and Condition Coverage. All machine commands displayed in the window were executed (100% statement coverage) but the condition if (e == 0) in source line 970 was always met. Thus the jump bne ("**b**ranch if **n**ot **e**qual") at address 0x1404 was never executed. TRACE32 draws attention to this fact with an "incomplete" message at the source code level and a "(branch) not taken" message at the machine code level.

Strictly speaking, TRACE32 did not perform a Condition Coverage Analysis in the case shown but a MC/DC (Modified Condition/Decision Coverage) Coverage Analysis. The highlighted condition if (e == 0) is so simple that condition coverage and MC/DC coverage actually coincide.

The lower part of the Figure 54 shows the summary of the coverage analysis.

5.6.3 Limits of Hardware-Based Tracing

In connection with runtime coverage, tracing in general—regardless of whether hardware- or software-based—and runtime measurement, is dependent on the test vectors used during the observation period. Unlike runtime measurement, the observation period for hardware-based tracing cannot be extended arbitrarily in order to increase test coverage.

Often, the hardware of the product under investigation will not have the tracing interface available, so a special development version of the product must be used if hardware-based tracing is to be utilized. The same applies to the processor if only special derivatives of the chip support tracing, such as AURIX with its emulation and production version devices that have already been mentioned in this context.

Even if emulation devices and special developer hardware are available, they are not usable in every environment. In vehicles, for example, the use of hardware-based tracing is either not possible or only with difficulty. Just think of a transmission control unit surrounded by oil in the transmission housing: even the smallest of holes for a cable cannot be implemented.

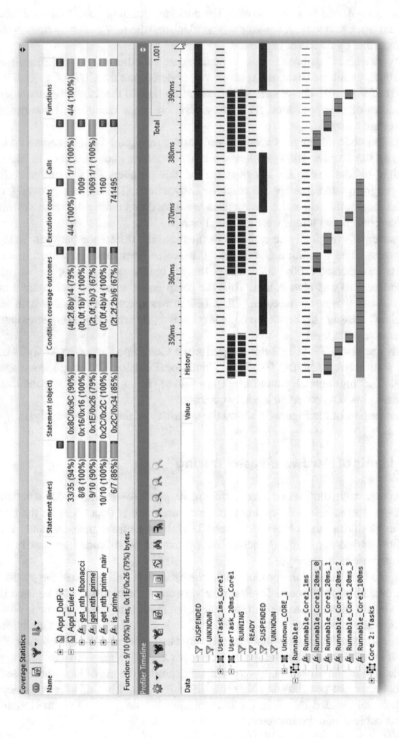

Figure 53   Screenshot from winIDEA [20], an IDE from iSYSTEM

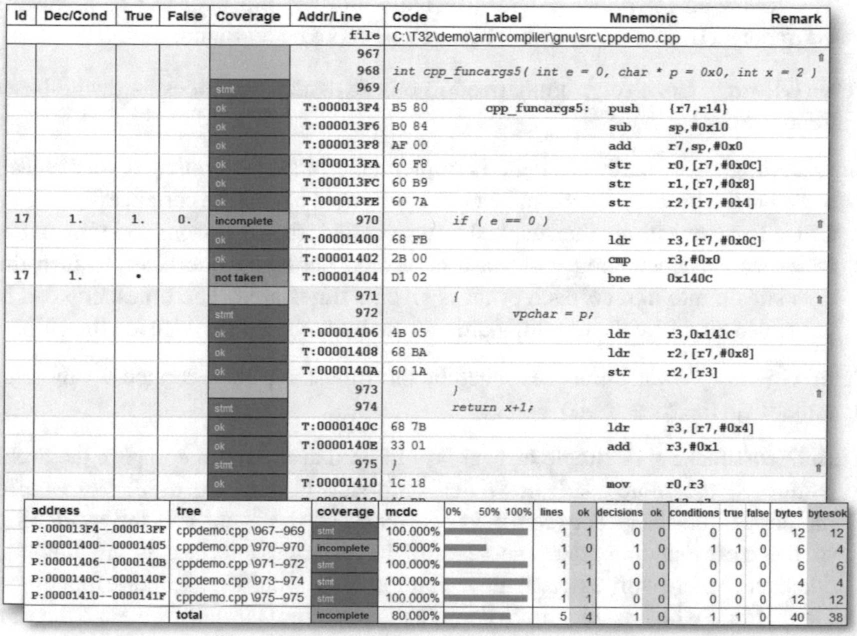

Id	Dec/Cond	True	False	Coverage	Addr/Line	Code	Label	Mnemonic	Remark
					file	C:\T32\demo\arm\compiler\gnu\src\cppdemo.cpp			
					967				↕
					968	int cpp_funcargs5(int e = 0, char * p = 0x0, int x = 2)			
				stmt	969	{			
				ok	T:000013F4	B5 80	cpp_funcargs5:	push	{r7,r14}
				ok	T:000013F6	B0 84		sub	sp,#0x10
				ok	T:000013F8	AF 00		add	r7,sp,#0x0
				ok	T:000013FA	60 F8		str	r0,[r7,#0x0C]
				ok	T:000013FC	60 B9		str	r1,[r7,#0x8]
				ok	T:000013FE	60 7A		str	r2,[r7,#0x4]
17	1.	1.	0.	incomplete	970	if (e == 0)			↕
				ok	T:00001400	68 FB		ldr	r3,[r7,#0x0C]
				ok	T:00001402	2B 00		cmp	r3,#0x0
17	1.	•		not taken	T:00001404	D1 02		bne	0x140C
					971	{			↕
				stmt	972		vpchar = p;		
				ok	T:00001406	4B 05		ldr	r3,0x141C
				ok	T:00001408	68 BA		ldr	r2,[r7,#0x8]
				ok	T:0000140A	60 1A		str	r2,[r3]
					973	}			
				stmt	974		return x+1;		↕
				ok	T:0000140C	68 7B		ldr	r3,[r7,#0x4]
				ok	T:0000140E	33 01		add	r3,#0x1
				stmt	975	}			↕
				ok	T:00001410	1C 18		mov	r0,r3

address	tree	coverage	mcdc	0% 50% 100%	lines	ok	decisions	ok	conditions	true	false	bytes	bytesok
P:000013F4--000013FF	cppdemo.cpp \967–969	stmt	100.000%		1	1	0	0	0	0	0	12	12
P:00001400--00001405	cppdemo.cpp \970–970	incomplete	50.000%		1	0	1	0	1	1	0	6	4
P:00001406--0000140B	cppdemo.cpp \971–972	stmt	100.000%		1	1	0	0	0	0	0	6	6
P:0000140C--0000140F	cppdemo.cpp \973–974	stmt	100.000%		1	1	0	0	0	0	0	4	4
P:00001410--0000141F	cppdemo.cpp \975–975	stmt	100.000%		1	1	0	0	0	0	0	12	12
	total	incomplete	80.000%		5	4	1	0	1	1	0	40	38

Figure 54   Screenshot of the TRACE32 IDE from Lauterbach [21]

In practice, their use is usually limited to the laboratory or, at most, to the HIL (Hardware In the Loop test system).

ORTI was already mentioned in the Section 5.6.2. Tracing the running task, i.e. the task that is currently in the Running state, is a great advantage over no OS awareness at all.

During development, however, the desire quickly arises to trace even more information regarding the scheduling. This is because the Ready states of the tasks are of interest, while the need for more detail, such as tracing the runnables, grows in importance.

It is becoming apparent that at least some of the desired features cannot be realized with hardware-based tracing alone. Today, hybrid approaches that mix hardware based tracing with instrumentation are used. However, as a result, the key advantage of hardware-based tracing, namely that the software for the analysis does not have to be modified, is lost.

5.6.4   Interview with Experts for Hardware-Based Tracing

Armin Stingl works for iSYSTEM [22] and Rudi Dienstbeck for Lauterbach [21]. Both companies develop and distribute debugging and trace tools for embedded software development. They also both operate worldwide and are active in standardization committees that aim to standardize tool interfaces for timing analysis.

The following interview provides insights into the practical use of hardware-based tracing, covers trends, and delivers some more background.

Peter Gliwa: Concisely summarized in one sentence: how does hardware-based tracing work?

Armin Stingl: A hardware logic is implemented in the chip that is very closely linked to the processor core and can be configured to capture certain events in the CPU. These events are then packed into so-called trace messages and transmitted to the connected tracing tool via a dedicated hardware interface. To turn this data stream into a trace, each event is given a timestamp. The timestamp can be generated either by the on-chip hardware or by the trace tool outside the chip.

Peter Gliwa: What influences whether the timestamp is generated on the chip rather than the trace tool?

Rudi Dienstbeck: A simple answer would be that the more complex the architecture is, the more likely it is that chip-side buffers are used. With buffer memories, the delay between the occurrence of the event and the availability of the corresponding trace message at the external interface is anything but constant. Thus, with external time stamps, you would make correspondingly large errors, which is why more complex architectures usually generate the time stamps themselves.

Peter Gliwa: Can you give some examples of common architectures and their implementations?

Armin Stingl: Both the NXP PowerPC architecture and the Renesas RH850 use Nexus. This is a standard that defines the hardware interface for tracing as well as the protocols that run over it. With Nexus, the time stamp comes from the trace tool, that is, it is generated externally.
The Infineon AURIX can be configured for both approaches. However, since the on-chip buffer memory is usually used, the timestamps generated by the AURIX are usually used. This also applies to the new ARM architectures and is the current trend in general.

Rudi Dienstbeck: It is important to understand in this context that the internal timestamps result in very high bandwidths.

Armin Stingl: Correct. We only use them when absolutely necessary.

Peter Gliwa: How wide is such a hardware timestamp?

Armin Stingl: Typically 32 bits.

Peter Gliwa: How does the trace tool become a tool for timing analysis?

Rudi Dienstbeck: From the traces it can be deduced, for example, when a function was executed. Interruptions, caused by interrupts or exceptions, can also be made visible. With all this information, the net runtimes of all functions can then be calculated in post-processing. The post-processing is then carried out conveniently by the trace tools.

Peter Gliwa: What are the core use cases? First in general terms, and then with regard to timing.

Armin Stingl: The most important aspect is certainly what I term 'advanced debugging', the debugging by means of program history analysis. When the software includes a failure (software bug), I have the ability to use the recorded program flow to determine how the failure occurred. From the flow trace I can see what code was executed before the error, i.e. 'where I came from'.

In other cases, I can use data tracing to determine, for example, how invalid access to protected data areas occurred. This allows me to quickly determine who the 'write culprit' is.

Peter Gliwa: A look at the program flow and its history is extremely helpful in itself. What about data? As a user, I would like to see the data relevant to the error case for each program step as well. Is this possible?

Rudi Dienstbeck: That depends on the processor. Some CPUs can also output the data, others not.

Peter Gliwa: What other uses does hardware-based tracing offer me?

Rudi Dienstbeck: One use case that is becoming increasingly important is code coverage.

Armin Stingl: In addition, data profiling should also be mentioned. This can answer questions such as: Which CPU accessed which shared memory areas and how often?

Up until now, these were use cases related more to the hardware-level. But hardware-based tracing can also help at higher levels of abstraction. State machines, for example, which are implemented by the software, can be made visible using this approach.

Peter Gliwa: Interesting, I was not aware of that at all! But again, back to the levels 'at the bottom'. What about the analysis of cache misses and pipeline bubbles?

Rudi Dienstbeck: I was just about to mention this as another use case. We actually do cache analysis for dedicated CPUs. Of course, this is already very time-consuming because we have to simulate the access history of the caches.

Peter Gliwa: So you reproduce precisely the caches and their logic, thereby deriving the cache content from the memory accesses? This goes far beyond hardware-based tracing.

Rudi Dienstbeck: Yes, exactly. That's how it works.

Peter Gliwa: I did not know that either. I have two further questions. First, if I start tracing at any time, I don't know what state the cache is in at that moment.

Do I have to wait until all cache lines have been evicted, i.e. the system has 'settled in'?

Rudi Dienstbeck: That is correct, and this is also true in principle for some of the runtime analyses we do. At the beginning of the tracing I don't know what state the system is in. As a rule, we do not start tracing at reset.

Peter Gliwa: Okay. Second question: What do you do if the CPU is using a random cache replacement strategy? Dice?

Rudi Dienstbeck: With Random Cache Replacement, the overwriting of cache lines cannot be traced in principle, which is why the cache analyses cannot be calculated exactly. However, Random Cache Replacement occurs rather rarely and therefore this limitation does not play a major role in practice.

Armin Stingl: Requests for pipeline analyses have to be turned down. It is simply not possible. What is also not possible, but is often requested, are bus load analyses.

Peter Gliwa: 'Bus' now in the sense of 'processor internal bus', like the crossbar, correct?

Armin Stingl: Yes, correct. Contentions such as access conflicts and their resulting delays in access, are particularly interesting. These sometimes have a massive impact on the latency of communication.

Peter Gliwa: This is one of the reasons why a purely static WCET analysis in a multi-core environment has no practical value: the theoretical worst-case assumption of maximum bus conflicts would be extremely high, but also extremely unlikely.
But, back to the nice processor features that could be conceivable. For example, the detection of access conflicts on the bus or at the interface to internal memories. Aren't you talking to the processor manufacturers and discussing future architectures with them to bring in such customer requests?

Rudi Dienstbeck/Armin Stingl: (both laugh)

Armin Stingl: It's difficult. We already talk to them, but usually relatively late when it comes to the debug and trace interfaces. To get new features into the chip, you would have to talk to the chip's architects much earlier. And that is not easy. How do you see it, Rudi?

Rudi Dienstbeck: We're already talking to them but new ideas and features ultimately cost silicon area and, in the end, it is the lower price that wins and new, helpful debug and trace features fall off the requirements list. At one chip manufacturer we were once very far into discussions but, at the end, it was not enough to result in any changes.

Armin Stingl: In my former working life I used to be a chip architect, and I was precisely the one who implemented the debug and trace units. There was always a fight between the project managers. They were under pressure from the customers. The customers are only willing to pay for application functionality but not for good analysis of the software that will later run on that chip.

The customer demands the ability to analyze their application when problems occur, but they do not want to pay for it in advance.

Rudi Dienstbeck: (laughs) That's it!

Armin Stingl: The architects are aware of the situation but their hands are tied. However, there are plenty of exceptions. For example, you can do a lot with the performance counters in the AURIX.

Peter Gliwa: Now to another topic, the bandwidth of the tracing interface. It is limited and I'm interested in what impact this has in practice.

Rudi Dienstbeck: As a user, I have to ask myself what I would like to trace. If too much is selected, the bandwidth may not be sufficient. The required bandwidth for a particular trace configuration depends on the software itself. Classic example: the idle loop of the operating system or the application itself, if no operating system is used. If this loop is very tight, a large number of trace messages are generated due to the frequent jumps and the bandwidth may not be sufficient. In this case, some CPUs allow you to exclude the idle loop from tracing.

Another very simple measure is to include lots of NOPs in the loop so that the idle loop still 'does nothing' but gets by with significantly fewer jumps, thus generating significantly fewer trace messages.

Peter Gliwa: Here we have a concrete tip for the users—thank you very much! How do I know, as a user, that the bandwidth was not sufficient? What is displayed in the user interface in such a case?

Rudi Dienstbeck: In this case we display a "Trace FIFO overflow".

Armin Stingl: To get a feeling for the topic of bandwidth: I'd say that with the current architectures you can run a full program flow trace for no more than six CPUs at the same time.

Rudi Dienstbeck: If you take the newest ARM CPUs, then three or four cores are enough to overload the tracing interface.

Peter Gliwa: Currently, autonomous driving is the big hype and the corresponding control units are being equipped with ever larger and greater numbers of cores. Doesn't it scare you that hardware based tracing doesn't scale well with the number of cores?

Rudi Dienstbeck: One has to limit tracing to the cores that are currently in the focus of observation. This is not new and the problem has existed for a few years with many multi-core processors.

The processor manufacturers react to this by increasing the bandwidth of the tracing interface. Some of them go out of the silicon at 10 GHz. The board designers then also have to cooperate accordingly. The lanes must all be cleanly shielded and of equal length, special RF connectors must be used, and so on.

Armin Stingl: This may sound a little outlandish, but it is definitely being done and used. Often, existing application infrastructure is used as well, especially for SoCs (System on Chip), such as an LVDS or PCI Express (PCIe) interface.

In general, it is becoming apparent that, in the future, you will have to trace in a more targeted manner, i.e. you will have to think a little more about what you want to trace.

Another aspect that is becoming increasingly important is that tool manufacturers, chip manufacturers, and software producers—for example, of operating systems—must coordinate their efforts more closely. Otherwise, there will be no efficient solution for tracing in the future. This is especially the case as it becomes more and more common for hardware-based tracing to be supplemented by instrumentation of the software. We already see this today. If the classic ORTI Traces are to be supplemented by the information as to which runnable is currently running, this is usually achieved using instrumentation.

Peter Gliwa: A better coordination of tool manufacturers, chip manufacturers, and software producers then not only delivers efficiency, it also makes it easier for the user to operate the system. Ideally, they can then set what they want to see at a high abstraction level, such as in the AUTOSAR Authoring Tool or in the Operating System Configurator.

Armin Stingl: And it saves costs. I then *do not* need the super fast tracing interface but get by with less bandwidth and, as a user, I still get to see what interests me.

Rudi Dienstbeck: Another thing I would like to mention: currently there are more and more chips where you don't even need to trace via external hardware. Instead, the data is written to a memory attached to the chip, for example DRAM. However, this does not resolve the bandwidth problem.

Peter Gliwa: Let us now turn our attention to the topic of timing analysis. Do you share the impression that timing analysis has gained in importance over the last few years? Does this also apply to tracing via hardware?

Rudi Dienstbeck: Absolutely 20 years ago, when OSEK came up, it was easy to calculate all the parameters of the scheduling and you could easily prove whether a task would always get its turn or not, whether it would meet its deadline. Today, such calculations are practically impossible with the complex systems in use due to multiple CPUs, multiple caches, pipelines, and so on. Implementations using AUTOSAR Adaptive Platform and POSIX systems in general make this even more complex. That means, for the code validation, you have to measure, to trace.

That is one point. The other concerns certification. More and more, our customers have to prove that the system behaves as specified, not only on paper but on the real target. This applies to both timing and code coverage. For example, Object

Code Coverage: no software tool in the world can do this for you; you need a hardware tracer.

Armin Stingl: I can only confirm that. Additionally, it should be mentioned that timing analysis is used more and more frequently in connection with Continuous Integration tools. The timing analysis of the software is simply called up via Jenkins after a nightly build.

Rudi, a comment on the two aspects you mentioned. If we now had experts in static code analysis in the round here, I'm sure they would immediately raise their finger and say, "Wait, we analyze both OSEK/AUTOSAR CP systems and AUTOSAR AP software."

Rudi Dienstbeck: Yes, they actually say that, but I am very skeptical about it.

Armin Stingl: What I see is that hybrid approaches are increasingly being used. The representatives of static analysis know that they can only analyze a part statically and need trace data for the complete image. By the way, this also applies to static code analysis.

Rudi Dienstbeck: Often, even the results of the measurement are transferred into the model.

Peter Gliwa: Regarding the keyword 'certification', I can immediately think of an argument that the representatives of static analysis—both at the code level and at the scheduling level—express. Namely, that only a model-based approach is worst-case capable because only it is independent of test vectors as well as the duration of traces, measurements, or simulation. How do you reply?

Rudi Dienstbeck: What is often done is what Armin called a hybrid approach. A worst case is calculated by static analysis that is then adjusted and measured with our trace tools. The measurement also includes interrupts and memory access conflicts, which the static analysis ignores.

Peter Gliwa: Tracing as a model checker, so to speak.

Rudi Dienstbeck: Yes, you could say that.

Armin Stingl: The theoretical approaches are usually too pessimistic for practical application.

Peter Gliwa: It's a pity that the static analysis tools always indicate only this safe upper bound and not the probability curve over the timing parameter just considered. If the probability of the WCET is lower than the probability that every single person on this earth will be struck by his own individual meteorite within 1 s, what practical value does this WCET have then?

Armin Stingl: Within a second? Yes, that can happen (laughs).

Rudi Dienstbeck: The possibility is given, however improbable it may be.

Peter Gliwa: It would be best to know the probability plotted over the CET and below a certain probability I simply say, "Quality target met, I can stop analyzing and optimizing."
If one is exclusively and strictly attached to this theoretical WCET and the WCRT, surely it is unhelpful in those areas where costs do matter.

Armin Stingl: Let's just say that we are not worried about becoming unemployed in the future. The real world is simply too complex for me to ignore it and rely solely on theoretical considerations.

Peter Gliwa: What other practical tips or recommendations can you provide to the reader?

Armin Stingl: My general recommendation is to think as early as possible about what will need to be tested and analyzed later. Just talk to the tool manufacturer of your choice to understand what is necessary, where the limits are, and what can be done to prepare. At the early phase of development it often costs nothing or very little to pave the way for the trouble-free use of tracing later on.

Rudi Dienstbeck: I can only support that. Just keep your options open.

Armin Stingl: When designing the hardware, don't just think about the debug port but also see if the processor offers a trace interface. It should be the task of every project manager to do this.

Peter Gliwa: That leads to a question that I have been asking myself for over 20 years: Why don't they design every ECU to include a single cheap LED for debugging purposes, at least until the C-sample? The automotive industry could have accelerated the debugging of software development in thousands of projects by days at a time.
Thank you very much for your time and your answers!

Rudi Dienstbeck studied computer science in Mannheim and has worked for Lauterbach since 1995. His area of responsibility has been supporting real-time operating systems in TRACE32 from the very beginning. He is co-author of the OSEK-ORTI standard and is an active member of AUTOSAR.
Lauterbach GmbH, founded in 1979, is the world's leading manufacturer of microprocessor development systems. Through very-close and long-standing cooper-
ations with all major semiconductor manufacturers, they are able to offer a debugger for new products as soon as they are introduced. Lauterbach is an independent and privately held company based in Höhenkirchen near Munich, Germany, with subsidiaries in Great Britain, France, Italy, Japan, Tunisia, China, and on the East and West Coast of the USA.

Armin Stingl joined iSYSTEM AG in 2013 and is responsible for the definition, validation, and market launch of novel debugging and testing tools. With over 20 years of professional experience he was involved in the development of several RISC cores. As a system engineer, he was responsible for architecture definition at well-known semiconductor manufacturers. In this capacity, he was involved in the development of on-chip debug and trace solutions for several processors.

iSYSTEM was founded in 1986 and is a privately owned company headquartered in Schwabhausen, near Munich, with subsidiaries in Slovenia and the USA. iSYSTEM offers embedded software development and test tools with a focus on customers in the automotive electronics industry.

5.7   Instrumentation-Based Tracing

Hardware-based instruction or flow traces usually target the lower levels of code. Tracing that is implemented using instrumentation of the software can also go down to individual machine instructions in terms of granularity, but the focus is more usually the scheduling level. The visualization of tasks and ISRs plays a decisive role here.

For millions of years, evolution has optimized the human brain to process large amounts of *visual* information in a short time. Our sense of smell is—well—present, our hearing is good, but our eyes and our ability to see are phenomenal. Man is a strongly visual living and thinking being. Irregularities in the landscape, groups of people, or in patterns 'catch the eye'. So whenever large amounts of data cannot be analyzed by means of simple rules or algorithms, it is a good idea to transform them into a graphical representation and leave the analysis to our powerful human brains.

A scheduling trace translates even complex runtime situations into a graphical form that our brain can handle very well. Irregularities and repetitions, as well as workload, distribution, and optimization potentials, are captured in fractions of a second and become almost tangible. The same data in a non-visual form, such as in naked unprocessed columns of numbers in the trace memory without translation into a visual form, are practically useless.

5.7.1   Basic Functionality and Workflow

Similar to the runtime measurement discussed in the Section 5.5, the software under investigation is supplemented by additional tracing software that records the events of interest. This can be done in two ways that are explained in more detail below.

Figure 55 compares the two approaches and also includes hardware-only tracing in the comparison. All elements that are used exclusively for tracing are shown in orange.

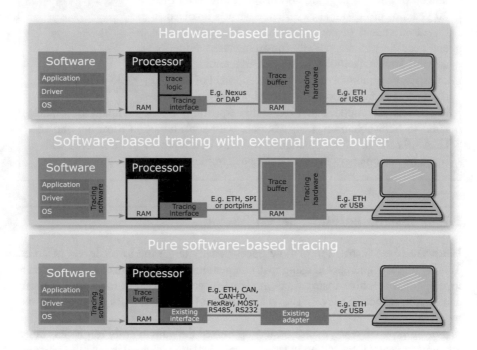

Figure 55   Three different tracing approaches with tracing elements in orange

5.7.1.1 Software-Based Tracing with External Trace Buffer

Although this approach is not a pure software solution, trace data acquisition is still based on instrumentation. The connection to the external trace hardware can be made via all conceivable interfaces, as long as these provide sufficient bandwidth. Port-pins, SPI, and Ethernet are classic interfaces for this approach.

The time stamps for 'software-based tracing with external trace memory' can either be generated by the tracing software itself and transmitted to the outside together with the other information, or they can be generated by external hardware. In the first case, the overhead resulting from the instrumentation will be higher. In the second case, the accuracy of the time stamps suffer as there is a variable amount of time (sometimes more, sometimes less) between the event occurring and the entry being made into the (external) trace memory.

The topic 'overhead' in conjunction with instrumentation will be discussed in more detail later.

For a few systems it will be possible to follow this approach without special hardware and to read the trace data directly into the PC and store it there. In most projects, however, this approach will fail due to the lack of real-time capability of the PC. Without an additional intermediate buffer in the embedded system, data will be lost.

5.7.1.2 Pure Software-Based Tracing

An alternative approach to external trace memory is to enter the trace information into a trace memory, which is implemented as a regular array in the processor's RAM. The trace information consists of information about the event and the timestamp.

The transfer of the contents of the trace memory for visualization and profiling is usually implemented using existing communication channels, i.e. those that are fed to the outside of the embedded system anyway. Of course, only a limited amount of bandwidth will be available for tracing on such channels before existing (application) communication is impacted.

5.7.1.3 The Instrumentation of the Software

In most cases, tracing that is based on the instrumentation of the software aims to capture the scheduling. But before this aspect is considered in more detail, the code introduced with the pin toggling measurement (see Section 5.5.1) should also be adapted for software-based tracing. The `main` function (see Listing 16 on page 105) remains untouched, as with the software-based runtime measurement, and only the files `observation.h` and `observation.c` change. These are shown in their new form in Listings 22 and 23.

Listing 22 `observation.h`: Tracing exclusively based on software

```
 1  #ifndef OBSERVATION_H_
 2  #define OBSERVATION_H_
 3
 4  #include <avr/io.h>
 5  #include <avr/interrupt.h>
 6
 7  #define NOF_TRACE_ENTRIES     (200u)
 8
 9  typedef enum {
10      START_EVENT,
11      STOP_EVENT
12  } info_t;
13
14  typedef struct {
15      info_t            info;
16      unsigned short timeStamp;
17  } event_t;
18
19  extern event_t traceBuffer[];
20  extern unsigned char traceIndex;
21
22  void InitObservation(void);
23
24  inline unsigned short SafeReadTCNT1(void)
```

```
25  {
26      unsigned char   sreg;
27      unsigned short retVal;
28
29      sreg   = SREG;   // save interrupt lock status
30      cli();           // disable all interrupts
31      retVal = TCNT1;  // read consistent 16 bit timer value
32      SREG   = sreg;   // restore interrupt lock status
33      return retVal;
34  }
35
36  inline void traceEvent(info_t info)
37  {
38      unsigned char sreg;
39      if (traceIndex < NOF_TRACE_ENTRIES) {
40          sreg = SREG; // save interrupt lock status
41          cli(); // disable all interrupts
42          traceBuffer[traceIndex].timeStamp = SafeReadTCNT1();
43          traceBuffer[traceIndex].info = info;
44          traceIndex++;
45          SREG = sreg; // restore interrupt lock status
46      }
47  }
48
49  inline void StartObservation(void)
50  {
51      traceEvent(START_EVENT);
52  }
53
54  inline void StopObservation(void)
55  {
56      traceEvent(STOP_EVENT);
57  }
58
59  #endif /* OBSERVATION_H_ */
```

Listing 23 observation.c: Software-only tracing

```
1  #include "Observation.h"
2
3  event_t traceBuffer[NOF_TRACE_ENTRIES];
4  unsigned char traceIndex;
5
6  void InitObservation(void)
7  {
8      // !!! careful !!! function relies on reset values
9      TCCR1B = (1 << CS10); // timer start running at full speed
10     traceIndex = 0;
11 }
```

As shown in the lower part "Pure software-based tracing" of Figure 55, the code uses part of the processor's own RAM to store the trace data. The example uses 200 entries. Each entry consists of a timestamp and information regarding *which* event was entered. In the case of the example, only two events are defined, specifically start and stop of the measurement, but many more events could be added.

But now back to the typical field of application for software-based tracing: the analysis of scheduling. The operating system as the component that organizes the scheduling at runtime is the best place for instrumentation. Some operating systems provide interfaces for instrumentation by default.

The tracing software is called via *hooks*, if the operating system—at least partially—is available in source code, or via *callouts* if the operating system is delivered in the form of object code (usually packaged as a function library). Hooks are macros ("`#define` ...") with the great advantage that they do not generate any overhead when not in use. As empty macros they are removed by the preprocessor before the actual compilation.

Callouts are regular function calls from the operating system. The called functions must be implemented by the user or the supplier of the tracing solution. If callouts are used, there are several possibilities, but all of them have their disadvantages.

An unconditional function call necessarily requires the implementation of the called functions, otherwise the linker will complain about unresolved externals (see Section 1.3.6 for background information on this error message). The functions are always called and therefore incur runtime, even if they are empty (i.e. no tracing should take place).

A call via function pointer solves the problem of unresolved links when tracing is not used, but now requires a check for each event to be traced as to whether tracing is active and whether the tracing software must be called. This one `if(...)` also incurs a runtime penalty. Furthermore, some coding guidelines prohibit the use of function pointers.

This leaves the last solution, which is the most efficient in terms of software, but complicates logistics. There are simply two variants of the operating system: one in which tracing including callouts is active, and one without any tracing callouts.

The Section 10.4 deals with a standardization of the interface between operating system and tracing tool. ARTI relieves the user of most of the worries regarding instrumentation. If both the operating system and the tracing solution support ARTI, the code generators take over the instrumentation.

Even if the operating system does not provide an interface for tracing, a generic interface for tracing can be created in most cases. 'Generic' in this context means an instrumentation implementation that does not require manual modification of the code at each task and interrupt. Such an approach would be extremely error-prone. If, for example, another task were to be added to the system at a later time, it would also have to be manually instrumented later. The risk is high that this will be forgotten. If the new task has a high priority, it would certainly interrupt other tasks at runtime, but would not be visible in the trace. The net runtimes of the interrupted tasks would be accordingly incorrectly calculated.

In addition to instrumenting the events relevant for scheduling, the user can add tracing to their own software components, such as the application software, as required. Thus they can capture interesting time periods (from-to), specific points in time (user events), or even capture user data to the trace memory. This is undertaken using the interface provided by the tracing tool.

Figure 56 shows such user-defined 'stopwatches', the term used in the T1 analysis suite [23], and are displayed as blue bars below the tasks. The colored arrows (three magenta ones and several green ones) in the trace are user-defined data flows. The start of an arrow corresponds to the 'send' event or writing of a variable, while the tip of the arrow marks the 'receive' event or reading from the variable.

T1 Traces for the representation of states of tasks and interrupts have already been used in several places in the book. In the Figures 20 and 21, the representation includes the runnables. All traces in the subsequent Chapter 6 were also created with T1.

5.7.1.4 Instrumentation at Runtime

If code is executed from RAM, tracing software can be 'injected' into the running system by means of code exchange. For example, if a function is to be instrumented at runtime, the first command of the function is replaced by a jump to the tracing software. The tracing software enters the start event in the circular buffer, executes the replaced command and the rest of the function, concluding by entering the end event in the trace buffer.

But, even if the code is in flash, there are still ways to instrument the code. Figure 57 shows how T1 uses the T1.flex component to instrument code at runtime, even if it is in flash. This dynamic runtime instrumentation has some very significant advantages.

1. The instrumentation actually takes place while the code is executed. This means that the code does not need to be compiled, linked, and flashed after instrumentation. This saves between minutes and hours, depending on the complexity of the project and the build process. The instrumentation at runtime is active almost immediately during operation.

   The manner in which we work in this respect is changing completely. We 'move through the code' while the software is running, instrument a function here and a loop there, and always acquire the current values of the most important timing parameters for the selection made.
2. The instrumented code itself does not change. Although the tracing software is added before and after execution, the instrumentation has no effect on the machine instructions of the instrumented code. The situation is different for instrumentation of the source code with subsequent compilation. Here the compiler will sometimes generate significantly different code and will not be able to execute some optimizations. This results in code that behaves differently in terms of runtime when compared to the non-instrumented version.
3. A positive side effect of the previous advantage is that instrumentation at runtime can really take place at the level of individual machine instructions.

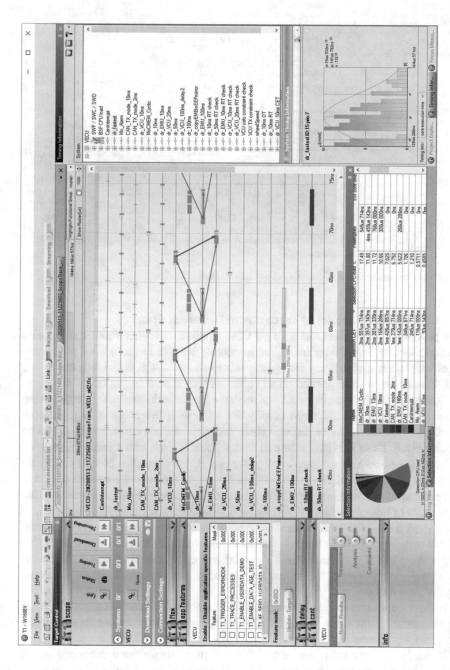

Figure 56 T1.scope trace with data flow information (colored arrows)

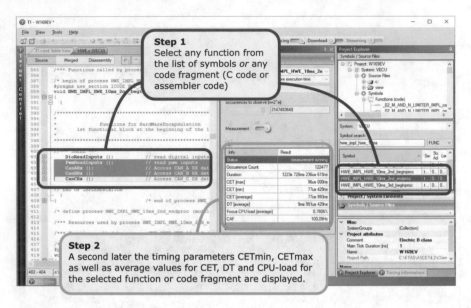

Figure 57   Instrumentation of code in flash at runtime using T1.flex

5.7.1.5 Synchronization of Traces from Different Sources

The previous explanations had a single CPU in mind, but what about tracing with multiple CPUs? Scheduling traces are extremely helpful in multi-core projects—especially when they display the synchronized traces of the individual cores—so that you can see what happened on all cores over a certain time period.

Whenever the traces to be synchronized have the same timebase, synchronization is usually quite simple. This is usually the case with the cores of a multi-core processor. It becomes more difficult if the traces of different processors are to be synchronized. Even if it is possible to find a point at time x that can be identified in the traces of both processor A and processor B, the time axes cannot simply be 'laid next to each other' from this point in time with the expectation that they will represent the same time in parallel from then on. The timebases are different and the trace times of both processors will drift away from each other (see also Section 6.4). With short traces, the error or drift may be so small that it can be safely ignored. This is certainly no longer the case for traces measuring minutes in length.

For short traces, a common time x must therefore be identified that occurs in all traces to be synchronized while, for longer traces, such synchronization times are required regularly.

It is comparatively simple if synchronization mechanisms such as the AUTOSAR "Synchronized Time-Base Manager" [24] are already in use for the applications on the various processors.

If this is not the case, a way must be found to detect an event occurring on all affected processors at the same time. Once this event occurs, a synchronization event must be entered in all traces to be synchronized as soon as possible after it. The description 'as soon as possible' is flexible and usually difficult to implement in practice. The next-best case is that the delay between the occurrence of the event and its entry in the trace memory is constant, known, or measurable. The delay can now be taken into account when synchronizing the traces.

Hint *At first glance, the remarks on the synchronization of different (trace) data sources seem to be theoretical in nature and at most relevant for the development of trace tools. However, it is worth considering and providing support for trace synchronization in the system design. If, for example, a port pin for a synchronization signal is provided on each of the processors, the traces can be synchronized easily, efficiently, and reliably later on. One of the processors is configured so that the pin is alternately set and reset when a timer overflows. For many microcontrollers this can be achieved by configuring the timer peripherals accordingly and without any interaction from the software. In addition, an interrupt is configured that triggers when the timer overflows. All other processors configure their port pins to an input that triggers an interrupt on edge changes. Both the ISR that triggers the overflow, and the edge detection ISRs associated with the port pin, enter a synchronization event into the respective local trace memory. This takes very little time and enables a very precise synchronization of the individual traces.*

Optionally, in addition to each synchronization event, a counter can be stored in the trace memory that is incremented by one for each event. This value then corresponds to a global time available on all connected processors.

If the provision of such port pins is planned early in the development process, the implementation is problem-free. However, if the desire for synchronized traces arises late in the course of the project—perhaps because problems occur in the interaction of the processors that must be investigated by means of tracing—a modification of the hardware is usually no longer possible. In such a case you have to settle for other, worse alternatives, such as synchronization using CAN messages.

5.7.2 Use Cases

5.7.2.1 Timing Analysis at the Code Level

At the code level, tracing can be used very effectively for profiling, i.e. the acquisition of timing parameters. This was already addressed in Section 5.2.2 and shown in Figure 40.

The situation regarding the net runtime during tracing is very similar to the situation during measurement as described in Section 5.3.1.2. Additionally, during tracing, all conceivable interruptions must be apparent from the trace data. This means that at least the execution of all tasks and ISRs must be recorded via corresponding events. This can be done either by instrumenting the tasks and ISRs directly, by instrumenting the operating system, or by using the hooks provided by the operating system.

This procedure allows you to select the events to be recorded. Depending on *what* is instrumented, the code level, scheduling level, or a mixture of both, is available. At the code level, for example, the body of a loop, a function, a runnable, a single task, or an ISR can be observed.

5.7.2.2 Timing Analysis at the Scheduling Level
If the instrumentation targets tracing the scheduling—i.e. the activations, starts, and terminations of tasks as well as the start and end of all interrupts are recorded— tracing is perfectly suited for the visualization and analysis of the scheduling.

The cause of a large portion of timing problems can be found at this level. Often, a quick look at the trace, with its display of tasks and interrupts, is enough to identify the cause of a problem that a team has been working on for weeks without such a visualization.

The same applies to the runtime optimization, something that is addressed in detail in Section 8.1. The graphical representation of the processing of tasks and interrupts over time quickly reveals potential for optimization.

5.7.2.3 Analysis of Data-Flow, Communication, and Synchronization
Up until now, we have only talked about tracing code at different levels. If data accesses (for example, writing/reading of global variables) or communication (sending/receiving) are instrumented, they become visible in the traces and can also be analyzed quantitatively. The frequency of access and the age of data, i.e. the time difference between the last write operation when reading data, should also be discussed.

Especially in multi-core projects, the communication and synchronization of the CPUs among themselves plays an important role. In Section 7, this topic is dealt with in more detail. However, it should be noted here that practically every multi-core project, sooner or later, gets into the situation that the behavior on the target hardware does not match the theoretical predictions or expectations. A look at the real system by means of tracing quickly reveals aspects that have not yet been considered, statically analyzed, or simulated.

5.7.3   Limits of Instrumentation-Based Tracing

5.7.3.1 Incomplete Test Vectors
What is true for code simulation, runtime measurement, and hardware-based tracing is also true for tracing based on software instrumentation. The results depend on the test vectors.

5.7.3.2 Overhead with Respect to Additional Runtime
The tracing software, essentially the additional code representing the instrumentation, requires runtime and thus generates a certain overhead. Whether this can be neglected, or whether it puts such a heavy load on the system that the actual application software can no longer be executed, depends on various factors: How

long does it take to enter a single event into the trace memory? How many events are recorded per second? At time-critical points, is the original code delayed to such an extent that timing requirements are violated?

With regard to overhead, the analogy with an oscilloscope can again be used. As soon as the probe of the oscilloscope is held against the circuit to be examined it is no longer the same circuit. The probe head has a certain capacitance, it may couple interference signals, EMC behavior changes, and so on. Whether all this is relevant and makes the measurement useless depends very much on the circuit and the measurement task. Is it a highly sensitive high frequency receiver in the gigahertz range, or just the logic signals to and from the CAN transceiver that are being examined? In the first case you would certainly have to give some thought to acquire meaningful results, while the second case is less problematic and the influence of the probe on the result can be neglected.

In practice, there is a wide range of very different tracing software available. Many in-house solutions, i.e. tracing and/or measuring tools developed by the users themselves, often consume several microseconds processing time per event and the CPU utilization required for tracing lies in the double-digit percentage range. Highly-optimized tracing software, on the other hand, requires as little as 60 ns per event for a second-generation AURIX; the resulting CPU utilization for tracing is thus less than 0.4 percent per CPU for most projects.

5.7.3.3 Overhead with Respect to Memory Consumption

Another aspect related to the keyword 'overhead' is the memory requirement of the tracing software. In addition to the program memory (flash), RAM is also used, especially for the trace memory if it is located in the RAM of the processor, as shown in the lower part of the Figure 55. To capture the desired events, the trace memory often has to be several kilobytes in size and not every embedded software project can provide that much RAM for tracing.

5.7.3.4 Interface of the Tracing Tool to the Outside World

The overhead of the tracing software itself is one thing, while the interface to the outside is another. If a dedicated hardware interface is required, this can sometimes only be achieved with special development version of the hardware. A pure software solution, on the other hand, requires additional bandwidth on the existing interfaces or buses. Depending on how the tracing software works, the demand for bandwidth can exceed the amount required for safe and reliable operation of the system. This danger exists with *streaming*, if all trace data is sent to the PC without interruption.

5.7.4 Interview with an Instrumentation-Based Tracing Expert

Fabian Durst, with whom the following interview was conducted, is, unlike the other interview partners, not a representative of a specific tool and he does not work for a tool provider. Fabian Durst is a user of timing tools and at his employer, Robert Bosch GmbH, he is involved in the provision of complete timing analysis infrastructure for the engine control division. One area of focus is software-based tracing.

Peter Gliwa: In short: How does software based tracing work?

Fabian Durst: With software-based tracing, the objects to be observed, such as functions or tasks, are enclosed by a wrapper in order to intercept the beginning and end of the processing of the object and to enter them into a trace memory as timing events together with a time stamp. When the trace memory is interpreted, it is then possible to reconstruct what happened, and when, at runtime.

Peter Gliwa: How exactly do the wrappers come into play? Are they part of the application or part of the operating system?

Fabian Durst: Usually it's a mixture of both. The timing events that affect scheduling come from the operating system and there are also application-specific events, where the tracing functions are called directly from the application code.

Peter Gliwa: Every year tens of millions of your engine control units are installed and a significant number of projects are currently under development.
You work across divisions to provide runtime analysis technology to individual ECU projects. What does that look like? Do you have a collection of expert tools, or is it an easy-to-use 'mass product'?

Fabian Durst: A mixture of both. Basically the tracing and the runtime measurement using T1 is provided by you, GLIWA. For this tool there is a default configuration that records, measures, and visualizes the scheduling, as well as providing information typical to the application. Furthermore, deeper insights can be attained by using different configurations or other tools, partly developed by us, that are used in addition.
The goal in development projects is to perform timing validation and simple timing analysis without a lot of time expenditure and without too much in-depth expert knowledge. In the case of more complex analysis tasks, or particularly stubborn timing problems, the timing experts from our team then support the projects.

Peter Gliwa: So you have a whole toolbox ready for timing analysis. What else is in there?

Fabian Durst: Of course hardware tracers, which we use both in the lab and now and then on the HIL. Also scheduling simulation and model-based procedures. We also have a number of self-developed tools.

Peter Gliwa: What do they do?

Fabian Durst: Various things. For example, there are various converters that convert timing related information from one format to another.

Peter Gliwa: A task that may take a back seat in the future, if ASAM ARTI prevails.

Fabian Durst: Quite possible. I'm curious to see how ARTI will develop. In addition to the converters we also support a whole range of other tools. Database-supported tools but also automated approaches that evaluate measurement and trace data and optimize the software according to the results.

Peter Gliwa: Do you also optimize symbols that are frequently accessed and put them into fast memory?

Fabian Durst: Exactly, this is an optimization approach that we pursue. We use measurement and trace data to determine the access frequency and the call frequency.

Peter Gliwa: How great is the potential there according to your experience?

Fabian Durst: Well, that's hard to say in general terms and depends very much on the original state of the software to be optimized and, of course, on the hardware used. However, the runtime gain of such an optimization can be quite decisive for the successful completion of a project.

Peter Gliwa: In any case it makes sense to automate this. Nobody wants to analyze thousands of symbols by hand and then assign them individually to the appropriate sections.
How do you bring all the tools together? Is there some kind of framework?

Fabian Durst: Format and methodology are mostly based on AMALTHEA[25]
...

Peter Gliwa: ...the now complete research project that provides a platform for embedded multi-core systems.

Fabian Durst: The exchange format is complex and allows us to bring together data from a wide variety of software and timing analysis tools. This includes call trees, runtimes, trace data, access frequencies, etc.

Peter Gliwa: What about static code analysis for determining WCET?

Fabian Durst: We can also take that into account, but we don't do it for the normal development path. Rather we save it for special analysis purposes on the platform. In any case, static WCET analysis is one of the tools that is operated by one of a few experts. The crucial point in the whole data collection and evaluation is to bring together the appropriate data, analyze it correctly, and then interpret it correctly. More important than the blind collection of vast amounts of data is the selective tracing of crucial situations. The traces or the timing data must be analyzed in connection with the respective driving situation.

Peter Gliwa: You also mentioned the scheduling simulation earlier. What do you use it for? The setting up of operating system configuration and timing layout for new projects? Or for optimizing task priorities and offsets?

Fabian Durst: We use the scheduling simulation at two development steps. The first is at the design of the basic system, although much is already based on defined standards. This applies mostly to the task priorities and offsets. The second is for scheduling simulation used in specific projects to, for example, efficiently distribute tasks across the different cores.

Peter Gliwa: And the input variables of the scheduling simulation—for example the minimum and maximum net runnable runtimes of the runnables—are determined by runtime measurement?

Fabian Durst: Correct.

Peter Gliwa: The core theme of the interview is, after all, tracing or measuring based on instrumentation. What are the most important use cases for you?

Fabian Durst: A very important use case is profiling, especially measuring net runtimes—i.e. core execution times—of tasks, interrupts, and runnables, as well as the response times of tasks. This is done automatically on the HILs and, in the vehicle, it is implemented by storing minimum and maximum values.

Peter Gliwa: You mean the recording of the minimum and maximum of a timing parameter?

Fabian Durst: Correct. These values are recorded by the system itself and read out at a later time. After all, there isn't always a developer sitting in the passenger seat with a computer on his lap constantly checking the timing.
Another important use case is the investigation of timing problems. I would call this *timing debugging*. It's very helpful because I can trigger tracing at various places in the software. This means that if the software detects implausible data during the plausibility check it stops tracing. In this way, T1 provides me with a trace that shows me exactly what was going on in the scheduling at the moment the problem occurred. This is supplemented by the ability to record application-specific data by means of 'user data events'. In the trace I can see the correlation between data and scheduling very clearly. Synchronization problems between the individual core in multi-core projects can be detected very quickly in this way.

Peter Gliwa: Can you give a concrete example from practice?

Fabian Durst: Recently we had a case where we observed maximum values in the drag indicators that we couldn't really explain. Had values of this magnitude appeared under normal driving conditions we would have observed massive problems, but the software ran without any noticeable difficulties. The cause was quickly found. We only needed to define a constraint—a limit value—for the corresponding timing parameter. The software did not even have to be modified for this. When the limit was exceeded during subsequent measurements, the tracing stopped. When analyzing the trace buffer contents it was immediately clear what was happening. In this particular case a high runtime occurred during shutdown, which we were then able to evaluate and optimize accordingly.

Peter Gliwa: So, a problem that wasn't—at least not in normal operation.

Fabian Durst: You could say that. But only by tracing and the contextual reference were we able to recognize this. The numbers from the measurement alone signaled 'we have a problem'.

Peter Gliwa: This fits quite well with what we discussed earlier: the timing data must be seen in the context of the driving situation. Or, more generally speaking, in conjunction with the basic state of the embedded system.

Fabian Durst: Exactly, I have to know the state of my system and ensure that those states occur during my measurements.

Peter Gliwa: How long does a single measurement like this take?

Fabian Durst: I can't give you a blanket answer on that. The duration of the measurement alone depends very much on the desired granularity and the any underlying problem. The range here goes from a few minutes to several hours. The measurements are performed by the project development team.

Peter Gliwa: This shows that the projects can work very autonomously.

Fabian Durst: Absolutely. We have created a framework, self-built, which supports the developers very well throughout their projects. The typical timing analysis tasks can be carried out without any problems. This framework is embedded in a process and the employees are trained accordingly.

Peter Gliwa: We talked earlier about use cases of software-based tracing. What about the use case 'runtime protection in series'? Does T1 operate and measure in the final software version on the road in production vehicles?

Fabian Durst: Well, yes, but only in a very limited way. Only some central timing parameters are monitored.

Peter Gliwa: Now for a completely different subject, the CPU load. Here it is very important which observation period is used for the calculation. How do you approach CPU load?

Fabian Durst: First of all, I am not a friend of CPU load. If someone tells me "My system has a CPU load of 92%" I first ask, "How did you measure that? Under what operating conditions? Was that a maximum value? What was the observation period?"
One issue is that this value depends on so many things and it can only insufficiently describe the complex system 'engine control' with all its associated timing requirements. This cannot be represented by a single number. In my opinion, the CPU load is clearly overrated.

Peter Gliwa: But it is so beautifully simple: one number and all is said!

Fabian Durst: This is the reason why it is so successful and persistently appears in all specification documents.

Peter Gliwa: I am curious to hear what you have to say about my views on the subject. I prefer *not* to demonize this very management-friendly CPU load measurement, preferring to mediate its use between managers and developers. We developers must be able to measure CPU load in a way that meets management expectations. This is not always easy. Of course, it cannot be that a system for which I measure 85% CPU load is overloaded and has failed task activations.

Fabian Durst: Difficult. When using the CPU load, you have to be very careful what kind of feeling you get and if it is the right one. Is the interval for which I determined it correctly chosen or are you lulling yourself into a false sense of security?

Peter Gliwa: Exactly. We agree on that. It is our job to understand the interrelationships, configure the system and measurement accordingly, and then perform the calculation correctly. If we don't succeed in doing this, we, as timing experts, have failed to a certain extent.

Fabian Durst: In the meantime, we have come to a different solution and have introduced a quantity that we call *system load*. This provides the maximum value of a whole range of parameters.

Peter Gliwa: This is interesting. So that's where the CPU loads of the cores and the bus loads come in?

Fabian Durst: Bus loads—not yet. Possibly in the future. But, besides CPU loads, task response times are also taken into account.

Peter Gliwa: I like that approach very much. Everyone can understand the term 'load', and the term 'system' indicates that the determined value is broader than that of just the load on a single core. The maximum value calculation is also useful. After all, a single timing requirement that is not met, or a single aspect that is problematic, is enough to turn the traffic light to red—to stick with the management view.

Fabian Durst: Correct. The introduction of a system load has proven to work well for us. By the way, it is not only timing parameters that are considered. The stack load is also taken into account.

Peter Gliwa: What I also like is that the transition from summarizing system load to a more detailed view is seamless. If I have a problem with the system load, the question immediately arises as to which of the parameters is the cause.

Fabian Durst: Correct. Instead of presenting long columns of numbers, I can represent it as a single number and, if it exceeds a certain threshold, I can go into detail at the correct location immediately.

Peter Gliwa: Back to tracing. Have you ever taken measurements or recorded traces yourself on the test track?

Fabian Durst: No, not yet.

Peter Gliwa: Really? That can't be true! That is something you have to do!

Fabian Durst: Well, we just have a well-functioning division of labor. The developers from the projects naturally also work on and in the vehicle during the timing analysis, but we timing experts are part of a cross-functional department.

Peter Gliwa: You don't know what you are missing. I remember once, with the computer on my lap, from the passenger seat, I used T1.delay to trigger runtime problems in a targeted manner, then used traces to analyze how the software reacts. That was a suspension project and it was all about tight turns all the time. After 20 min I felt terribly sick.

Fabian Durst: Sounds tempting.

Peter Gliwa: But seriously: I have often observed that even those responsible for tools in larger companies all too seldom experience, observe, and evaluate the use of the tools they are responsible for. After all, the users are basically their customers, even if they are of course their colleagues.

Fabian Durst: That's correct. It also makes sense the other way around. That the 'customers' come to us and understand the flood of requirements we have to meet. You learn a lot from each other when you sit together.

Peter Gliwa: Speaking of 'sitting together', how is cooperation when it comes to timing with your clients? After all, they are the ones feeding timing requirements or requirements regarding tracing and measurement technology into your system via their specifications.

Fabian Durst: That's correct. It's relatively easy for us because the timing analysis that is required is something we can cover in most cases with our existing setup. Dare I suggest that we are pioneers in this respect. We've put a lot of effort into it but it was worth it.

Peter Gliwa: Does it often happen that the customer wants to be put in the position to be able to measure and trace themselves? For example, in order to be able to analyze their own software components with regard to timing?

Fabian Durst: This is common practice and is supported by default. Customers can also trace and measure their own software elements with T1.

Peter Gliwa: Then I'll keep my fingers crossed for you to continue to be so successful with your self-developed analysis infrastructure. Thank you very much for your time!

Fabian Durst: With pleasure.

Fabian Durst studied electrical engineering at the Esslingen University of Applied Sciences until 2013. Both during his studies and during his previous vocational training as an electronics engineer for devices and systems, he maintained close ties with the Stuttgart-based technology company Robert Bosch, to which he returned after successfully completing his studies. Since then, Fabian Durst has been working in the central software development department for control units and is involved in the management of software resources in embedded systems. The focus of his activities is the development of measurement and validation concepts for software execution time in real-time systems and their tool support. Furthermore, he is responsible for the design and optimization of dynamic software architectures, the acquisition and evaluation of the computing power of microcontrollers, and the design and optimization of development processes in the context of resource management.

5.8   Scheduling Simulation

In the code simulation described in Section 5.4, the CPU was simulated as it executed a program. The *scheduling simulation* now simulates the operating system and interrupt logic as they orchestrate the execution of tasks and ISRs.

In the overview diagram of timing analysis techniques (Figure 39 on page 84), scheduling simulation is located a few levels higher than code simulation. This is because it is less detailed. Specifically, it rarely includes objects below the runnables.

5.8.1   Basic Functionality and Workflow

To start with, let's look at the workflow when using scheduling simulation (see Figure 58). Firstly, the simulation must be configured for a specific scheduling approach. Perhaps less precise but more clear: the operating system that is to be used for the simulation must be chosen.

Next the tasks and interrupts are created and the parameters relevant for scheduling are defined. The most important parameter is priority, while the others are multiple-task activation (see Section 3.2.4.1 on page 44) or the 'preemptable' setting (see Section 3.2.4.3). This completes the static aspects of the project configuration. However, the dynamic aspects are still missing. These are the minimum and maximum execution time for each task and for each ISR. The simulator also still has to be informed about the pattern according to which the tasks are activated and

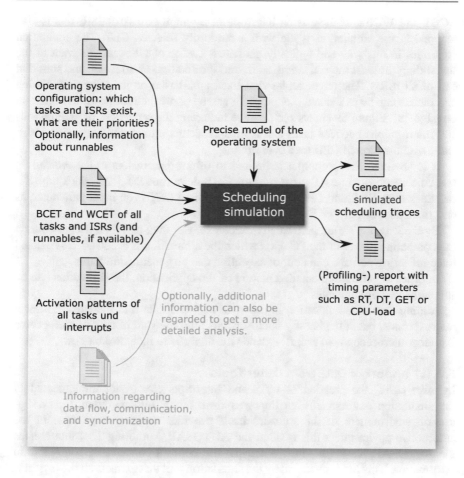

Figure 58 Input and output of the scheduling simulation

the interrupts are triggered. For the periodic tasks this is defined using their period and offset.

With the interrupts it is a little more complicated. One approach that has become established is to describe the occurrence of interrupts using *patterns of activation*. These were explained in the Section 4.2.4.

The last configuration parameter to be provided to the simulator specifies which statistical distribution is to be assumed for the runtimes between BCET and WCET such as, for example, Gaussian or uniform distribution.

Now that all inputs for the simulation are described, we can look at how the scheduling simulation functions.

The tasks are activated and interrupts are triggered in the simulation according to the activation patterns. For each simulated execution of a task or ISR, a runtime (CET) is now randomly determined according to the specified distribution and the

BCET and WCET. If there is an interruption by another task or ISR, this is also mapped in the simulation, again with a randomly selected CET. The simulation continues in this way and typically generates a trace of all events relevant to the scheduling (at least the activation, start, and termination of all tasks, and start and end of all ISRs). The trace can be visualized and, based on the trace data, timing parameters can be calculated (i.e. profiling can be performed as shown in Figure 40 on page 88). Figure 59 shows such a trace including the sequences of preemptions and interruptions between tasks and interrupts, status information regarding tasks, and a visualization of CPU load over time.

It is obvious that the output is limited to timing parameters at the scheduling level, i.e. response time RT, gross runtime GET, delta time DT, jitter, slack time ST, net slack time NST, and CPU load. The net runtime CET is an input parameter, as described above.

By the way, the simulation usually executes faster than real time. Depending on the computing power of the PC upon which the simulation is executed, a simulated trace reflecting perhaps one hour of execution can be generated in 5 min.

In order to get a more detailed picture of the application, the simulation can be supplemented by runnables.

Adding data communication in the style of 'send data d at the earliest after x microseconds, but at the latest y microseconds after the start of task A, and receive it after z microseconds in task B' extends the analysis to include data flows.

5.8.1.1 Import of Data from Other Tools
In most cases, the creation of tasks and interrupts etc. is not implemented in the simulation environment but imported from other sources. Often, the config-uration environment of the software itself provides the necessary data. In the automotive environment this is often the AUTOSAR Authoring Tool that creates the AUTOSAR XML (or short: ARXML) file with all conceivable configuration information. This file is then read by the simulation tool that extracts all the (system) configuration parameters required for the simulation. All that remains is to specify the dynamic input variables (BCETs, WCETs, and activation patterns).

However, even these dynamic parameters can be generated by tools. Tracing tools, for example, can provide all static as well as dynamic parameters so that the simulation can be started directly after importing the data.

A third approach is to use the simulation tool to read in trace data, interpret it, and then derive all the necessary parameters from it.

5.8.2 Use Cases

The scheduling simulation requires neither executable software nor ready-to-use hardware. With it, a timing layout, or timing concept, can be developed, analyzed, optimized, and even secured at an early project phase. In this approach the BCETs and WCETs of the tasks and ISRs can be understood as *budgets*. What does this mean? In general, the term 'budget' stands for a fixed amount of money that can be

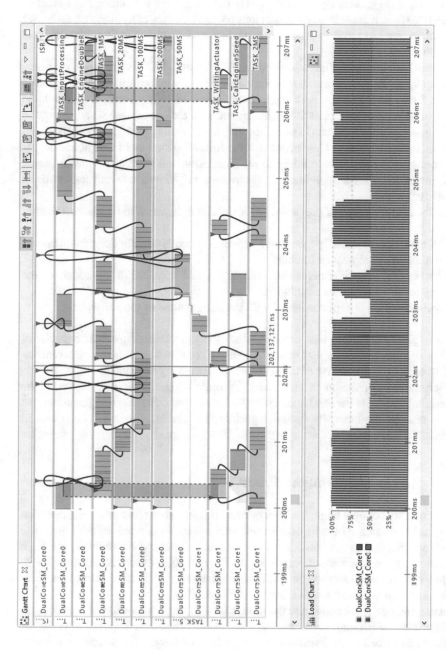

Figure 59 Screenshot of a trace generated with the TA Tool Suite [26]

spent for a specific purpose. In the case of timing, a budget is a WCET that may be spent on a task, ISR, or runnable.

So, where do these budgets come from in the scheduling simulation? They can be taken from the measured values of previous projects, or they can simply be estimates or assumptions. At an early project phase it is not so crucial to define these budgets very precisely.

Even if the entire configuration and the BCETs and WCETs are based on many assumptions, the simulation can be carried out and you can quickly see whether or not the system will run with the assumptions made. Any configuration related to timing—such as task priorities, offsets, budgets, etc.—can now be adjusted and optimized until the simulated system no longer has timing problems and meets all the timing requirements. This means that the following statement will hold true: "If we configure the system according to the simulation and if all components (tasks, ISR, runnables) stick to their budgets, the system will show a safe and stable timing behavior."

In the further course of the project, therefore, it only needs to be ensured that:

1. All components actually meet their budgets (which is usually done by means of runtime measurement or tracing).
2. Further timing requirements that arise or become apparent in the course of the project are added to the simulation.
3. The simulation is kept consistent with the current state of the software.

If these points are taken into account, scheduling simulation provides a powerful timing analysis technique that makes a decisive contribution to being, and remaining, in control of the timing of the software from the very beginning.

Scheduling simulation is also ideal for optimizing timing parameters at the scheduling level. On the one hand, manually by viewing and analyzing the simulated traces and, on the other, by optimizing algorithms, which Section 5.10 covers in detail.

5.8.3   Limits of Scheduling Simulation

If you quickly set up a configuration, or copy it from other tools or traces, and then start the simulation, you will usually find that the traces created reflect unrealistic situations. Similar to Section 5.3.3.5 with respect to static code analysis, there are also mutually exclusive tasks, interrupts, and/or runnables in scheduling simulation. Analogous to the annotation in static code analysis, the simulation must now be refined to such an extent that it delivers usable results.

Time and again, scheduling simulation is used in projects to analyze, understand, and solve acute (timing) problems. As valuable as scheduling simulation is for the use case described in the previous section, it is quite simply the wrong method for finding the causes of timing problems.

Here is a striking illustration of this. If a patient comes to the doctor with a pain in the chest, it is of little help if the doctor takes model of a heart from the cupboard, examines it, and then thinks about what could be causing the pain. He must examine the patient's *actual* heart by, for example, means of an ECG. The ECG provides insights into the actual function of the heart with a visualization that is very helpful for the doctor.

The analogy to tracing is obvious.

5.8.4 Interview with a Scheduling Simulation Expert

The topic 'scheduling simulation' will also be rounded off by an interview with an expert in this field. The experiences described in the interview relate to the TA Tool Suite [26] from Vector Informatik GmbH, where Dr. Michael Deubzer is the head of the product management team for the TA Tool Suite.

Peter Gliwa: Concisely summarized in one sentence: How does scheduling simulation work?

Dr. Michael Deubzer: Based on models of runtime behavior of the software, the hardware, and the scheduling approach, the most complete picture of the real-time behavior of the system is generated. Thus, the focus is not only on the worst case times but also on the 'normal' behavior with its average times.

Peter Gliwa: What are the central use cases?

Dr. Michael Deubzer: There are two main use cases for scheduling simulation: first, software architecture design and, second, software integration.
First, the software architecture design. Most projects are not created on a greenfield site but are derived from a platform. This platform contains the basic software and a basic framework of tasks and interrupts. By means of scheduling simulation, this basic framework and also the interaction of the basic-software components with each other can now be optimized. When deriving a project, the application software is added to the platform and the interaction of all tasks, interrupts, and communication is analyzed and optimized.
Even the structure of the software can still be modified and optimized within certain limits, such as the refinement of software components in runnables for example. Classically, synchronization mechanisms can also be designed and refined using scheduling simulation.
Scheduling simulation significantly speeds up the software architecture design process because it does not require the software to be built, flashed, and tested to verify a change. Instead, simulation is performed on the PC.
Now to software integration. This is about the optimal placement of the software. Software components are 'mapped' to OS Applications, runnables to tasks, and tasks to processor cores. Here, too, there is a lot of potential for optimization— especially when dynamic aspects play a role, such as mode-dependent executions of runnables or data flows.

By the way, on the keyword 'data flows': with scheduling simulation, whole chains of events can be considered and thus, for example, the age of data or buffer sizes can be determined.

Peter Gliwa: Functional simulation plays a central role in development at the functional level. Why is it that scheduling simulation is rather seldom found in system architecture?

Dr. Michael Deubzer: In isolated [automotive] function development, the subject of timing is already considered, such as in connection with sampling rates, for example. Unfortunately, for the entire ECU system, there is too often the assumption that "the basic software will fix it" or "we can check compliance with the timing requirements later". The effects of scheduling on data flows, event chains, etc., are of course disregarded in this approach—together with a corresponding collection of nasty surprises during commissioning.

If timing problems are determined, there is suddenly great panic and, in a 'firefighting' effort, they try to build a simulation. These usually take the approach of first tracing to understand the problem, then finding a solution at a higher level by means of scheduling simulation.

Peter Gliwa: At what timing resolution do your customers use scheduling simulation? Is it finished at the runnables level or are the called functions modeled as well?

Dr. Michael Deubzer: This is a good and frequently asked question, but the answer depends on the dynamic architecture of the system.

If, for example, when taking the AUTOSAR Classic approach, the application functions are called directly within the task and the execution logic (which may, for example, be dependent on an event or an application mode) occurs at task level, it is sufficient to model runnables with their runtime, communication, and synchronization behavior.

If, on the other hand, 'scheduling' functions are called within a task, which in turn call other functions according to a specific logic, then it makes sense to also map this logic.

This is the only way to start architecture optimization and improve the call behavior of these functions.

Peter Gliwa: What about at the upper limit of the 'timing resolution scale'? Are whole networks simulated?

Dr. Michael Deubzer: In the context of networks, timing issues often revolve around the end-to-end runtime of event chains. This can be, for example, the data flow starting from a sensor, through several ECUs and network media, to the actuator.

However, in many projects we have found that the analysis and optimization of such event chains lies either within the bounds of the ECU or in the area of network configuration. Depending on the developer's focus, appropriate timing tools are then also used for ECUs or networks.

In very early phases of vehicle-system design, however, scheduling simulation is not really used. In this phase, the relevant software components are often not yet completely defined and therefore their runtime behavior is not known. In such situation, where data is unknown, it is often more effective to achieve the goal by means of static calculations.

Peter Gliwa: Do many customers use scheduling simulation as part of their automated timing verification process? What are some of the approaches that have proven themselves and can be recommended?

Dr. Michael Deubzer: When it comes to the right side of the V of the V-Model, timing verification on the ECU complements scheduling simulation very well. In simulation, a wide variety of execution scenarios can be quickly executed and analyzed. This is especially the case for situations that occur very rarely in reality or are very difficult to reproduce. In addition, possible weak points in the dynamic architecture can be quickly identified.

For complete validation, however, it is necessary to check the system on the real control unit. By means of profiling, the critical scenarios can then be examined exactly under real environmental conditions and any simplifications made by the simulation can be examined in detail.

Peter Gliwa: Was there a practical experience with a certain 'aha effect' that you can share with us?

Dr. Michael Deubzer: So, when you look at multi-core systems, there are actually always very impressive 'aha' moments. One example is when the net runtime of a runnable is optimized on one core and then, as a result, the response time of a task on another core increases. Indirect dependencies between the cores due to shared resources are often overlooked in the optimization.

Peter Gliwa: What would you like to give the readers as a recommendation or practical tip?

Dr. Michael Deubzer: What has proven to be very useful in practice is the creation and maintenance of a runtime database that can later be used for scheduling simulation. This systematically records the net runtimes of runnables together with the processor used, the compilers, and other boundary conditions. This is done by means of profiling, either by measurement or tracing. With data obtained in this way, future projects can then be examined and evaluated by means of scheduling simulation. Example: for a new generation of ECUs, a supplier receives a specification sheet describing the desired functionality. The supplier can map the majority of the functionality with his existing software components. Since his runtime database contains extensive data on all runnables of the necessary software components, they can 'virtually assemble' and examine the desired ECU in a simulation. This answers whether the less expensive processor X is sufficient, or whether processor Y will have to be used in order to avoid a bottleneck in the runtime. Long before hardware and software for the ECU exist in a first version, reliable runtime investigations can be carried out.

Peter Gliwa: I can only agree with this recommendation. Anyone that has their timing view so well formed and under control not only saves on hardware costs, they can also expect to suffer from significantly fewer of the 'nasty surprises' you mentioned at the later stages of the project.

Thanks a lot!

Michael Deubzer received his doctorate in 2011 in the field of scheduling procedures for real-time systems with multi-core processors at the TU Munich. After his research work, he co-founded the Regensburg-based start-up Timing Architects, which developed the TA Tool Suite and distributed it worldwide.

The TA Tool Suite is used in the automotive industry to make the timing behavior of multi-core technology more controllable by means of model-based simulation procedures and trace analysis.

Seven years after its foundation, the 40-strong team of Timing Architects merged with Vector Informatik GmbH and has since continued product development within the framework of an AUTOSAR strategy. At Vector, he now heads up the product management team for the TA Tool Suite and other tools for optimizing software integration in AUTOSAR ECUs.

5.9   Static Scheduling Analysis

Static code analysis is to the code level as static scheduling analysis is to the scheduling level. It is a mechanism to determine worst case timing parameters following a 'mathematical approach' without recourse to simulation, measurement, or tracing. The timing parameters are, of course, the timing parameters at the scheduling level, especially the response time.

Static scheduling analysis can also be carried out at the communication level. For example, communication over the CAN bus can be verified and optimized in such a way that, although a bus load significantly higher than the widely used 40% is attained, it can be guaranteed that all messages are transmitted within their deadline.

The following anecdote shows that this is possible in practice and is anything but new. After Ford took over Volvo in 1999, there were a lot of meetings between the developers of both companies to exchange knowledge. Volvo had already started some time ago to validate the CAN communication of its vehicles by means of 'Schedulability Analysis' (an aspect of static scheduling analysis). This made it possible to ensure during the design phase that all messages would always arrive on time despite the high utilization of the buses.

When asked by a Ford network expert how Volvo conducts stress tests to validate communication, the Volvo expert replied, "Not at all. We design our networks from the outset so that no messages are lost."

Thereupon he was accused of lying and expelled from the room. Obviously not all developers were familiar with static scheduling analysis at that time.

5.9.1   Basic Functionality and Workflow

In order to define the concept of static scheduling analysis in principle, we will take a look at the formula for the most important timing parameter at the scheduling level, the response time (RT). Formula 16 is known as *Response Time Analysis (RTA)*.

$$
RT_i = \underbrace{J_i}_{\text{Jitter of task } i} + \underbrace{CET_i}_{\text{CET of task } i} + \underbrace{\sum_{j \in hp(i)} CET_j \cdot \overbrace{\left\lceil \frac{\overbrace{J_j + RT_i}^{\text{Observation interval}}}{PER_{0,j}} \right\rceil}^{\text{Number of preemptions}}}_{\text{Delay due to preemptions}} \leq \underbrace{DL_i}_{Deadline}
$$

$$(16)$$

The formula states what becomes obvious when looking at the trace of a concrete runtime situation, such as the one shown in Figure 28. The response time of a task T_i is composed of the absolute jitter J_i, the net runtime CET_i of the task, plus any delays that occur during the response time. Delays include those that fall within the initial pending time (IPT) as well as all interruptions after task T_i has been started. Of course, only tasks with a higher priority than the task T_i can cause delays. This group of tasks is described by $hp(i)$.

The number of interruptions of the task T_i by a task T_j is calculated by the quotient of the sum of the absolute jitter J_j of the interrupting task T_j and the response time RT_i sought, and the period $PER_{0,j}$ of the interrupting task T_j. The quotient usually provides a fractional number, while the corresponding number of interruptions is determined by the next largest integer. This is obtained by the *ceiling* function $\lceil x \rceil$ which 'rounds up'. To illustrate the ceiling function, here are a few examples: $\lceil 4.2 \rceil = 5$, $\lceil 1.9 \rceil = 2$, $\lceil 6 \rceil = 6$.

Two important aspects have to be taken into account. Firstly, the RTA does not consider *any* instance of the tasks T_i but the situation under which the WCRT (Worst Case Response Time) occurs: $RT_i = WCRT_i$. Secondly, the formula only includes the WCETs (Worst Case Execution Times). $CET_i = WCET_i$ and $CET_j = WCET_j$.

Additionally the deadline DL_i is specified, which can be understood as the maximum response time allowed.

Similar to a differential equation, the value RT_i that is sought appears on both sides of the equals sign and you cannot simply solve the equation for this value. As a result, solving the equation and finding RT_i is rather complicated—although the formula already makes restrictions which makes it unsuitable for real projects. It can only be applied to systems where all tasks are cyclical tasks with an offset of zero. If offsets are also to be considered, the analysis becomes much more complex. In addition, the formula assumes that the operating system has zero overhead—i.e. no time is required for a task switch—and that the tasks do not use any blocking mechanisms such as interrupt locks or the Priority Ceiling Protocol.

As mentioned at the beginning, the purpose of static scheduling analysis is usually to determine the worst case values of timing parameters, i.e. their maximum or minimum value. The art to developing a static scheduling analysis algorithm is to find an approximation method that can reliably provide the worst case, or a sufficiently good approximation to it. On the other hand, the algorithm should also work as fast as possible so that the analysis requires as little time as possible to execute. The analysis of more complex systems can quickly take several hours, even with sophisticated algorithms.

The analysis time plays an especially decisive role when optimization algorithms have to analyze many different configurations. In Section 5.10 this approach will be discussed in more detail.

What is the process when using static scheduling analysis, and what does the workflow look like?

Input data and output data are essentially the same as the scheduling simulation, which is why Figure 58 in the scheduling simulation section and Figure 60 for static scheduling analysis are very similar.

Unlike simulations, which execute until they are stopped, static scheduling analysis executes until the results are calculated. Instead of generating simulated traces of any length in the scheduling simulation, the static scheduling analysis can generate constructed traces of worst case scenarios. These are usually very short and show how a particular worst case scenario occurs. Static scheduling analysis is not suitable for determining and optimizing the *typical* system behavior.

Instead of simulated profiling data, the user receives the worst case timing parameters they are looking for.

5.9.2   Use Cases

Whenever worst case values have to be guaranteed at the communication level or scheduling level, static scheduling analysis is the best choice. The results are determined using formal methods and are therefore independent of test vectors and the duration of measurement, tracing, or simulation.

Static scheduling analysis can deliver very good results in practice when it comes to timing verification at the communication level, as the application at Volvo described above has shown.

Just as with scheduling simulation, neither executable software nor ready-to-use hardware is required. This makes it possible to verify the design and configuration with respect to timing at an early project phase.

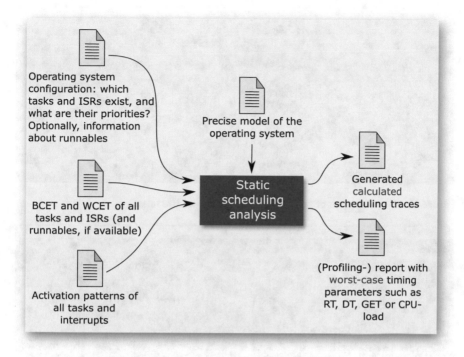

Figure 60 Input and output of the static scheduling analysis

But static scheduling analysis is not only suitable for securing a timing design, an OS configuration. It can also provide valuable assistance in optimizing timing parameters at the scheduling level. In Section 5.10 this will be covered in more detail.

5.9.2.1 Tool Example: chronVAL by INCHRON

Figure 61 shows the evaluation of a WCRT analysis for a single core software implementation with three tasks (T_20ms, T_10ms and T_5ms) and two interrupts (CAN and ISR_Sensor). chronVAL [27] from the INCHRON was used here. For each task there is a deadline that corresponds to the period of the task, such as $DL = 20ms$ for the task T_20ms. In the output, the tasks are sorted by descending period duration, resulting in task T_20ms appearing at the top. For each task and interrupt, two bars are visible: one for the WCRT and one for the BCRT. Each bar is further divided into a blue and a yellow section. The blue section illustrates the CET, i.e. the time resulting from the execution of the affected task itself. The yellow section represents the time resulting from interruptions or initial delays. In the first case (as shown in Figure 61 above), all tasks can be made to meet their deadlines under all circumstances. For task T_20ms, however, there is not much headroom left, as its WCRT is 19.98 ms.

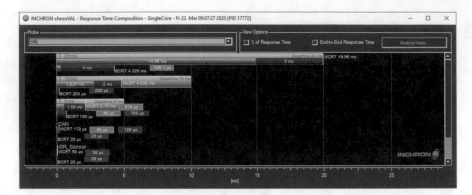

Figure 61 WCRT analysis of a system with three tasks and two interrupts

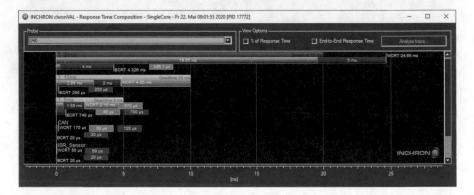

Figure 62 Just 1% more CET for task T_5ms results in a significant violation of the deadline of task T_20ms

If the execution time of the task T_5ms is now increased by only 10 μs, the WCRT of the task T_20ms increases abruptly to 24.65 ms, as Figure 62 reveals. The deadline is no longer met, which is graphically highlighted by the now red coloring of the deadline. The minimal increase in the execution time of the task T_5ms causes a further interruption of the task T_20ms and thus the step increase in time. The relationships between the different time periods in scheduling are not linear and, thus, a minimal change in a single input variable can have an unexpectedly large effect on the result. One might consider this to be a sort of butterfly effect [28].

Static scheduling analysis can also be used in later project phases to take into account corner cases, i.e. cases that occur extremely rarely and are therefore difficult to reproduce using simulations or tests.

5.9.3   Limits of Static Scheduling Analysis

Just as the use cases of static scheduling analysis partly overlap with those of scheduling simulation, both approaches share some of the same limitations (see also Section 5.8.3).

If, for example, different application modes with mutually exclusive tasks, interrupts, and/or runnables are implemented in the software, these must also be modeled in order to avoid unrealistically high results in the analysis. The same was already concluded with regard to simulation.

Like scheduling simulation, static scheduling analysis is often used as a means of problem analysis to find the causes of timing problems. However, in order to first understand the problem and isolate the cause, tracing is a much better approach. Whether this is undertaken by instrumentation of the software or by hardware-based tracing is of secondary importance.

The following interview shows that this view is not necessarily shared.

5.9.4   Interview with a Static Scheduling Analysis Expert

Dr. Ralf Münzenberger is a managing partner of INCHRON AG [29], a company that offers products for scheduling simulation as well as for static scheduling analysis. In the following interview, Dr. Münzenberger answers the questions mainly from the point of view of static scheduling analysis.

Peter Gliwa: Concisely summarized in one sentence: How does static scheduling analysis work?

Dr. Ralf Münzenberger: The starting point is a model, that we name a timing model, that contains all modeling elements required to perform timing analysis. This includes information about the executable units, i.e. the tasks, the interrupt service routines and—if available—the runnables. Specifically, we need their scheduling parameters such as priorities, processing time requirements, and an idea of when interrupts are triggered and when tasks are activated. For example, do the activations occur periodically or sporadically?
From this information we then calculate timing parameters of executable units at the scheduling level, the most important of which is the response time.

Peter Gliwa: Which other timing parameters play a role and can be calculated?

Dr. Ralf Münzenberger: The delta time of periodic executable units shows how large the deviation from the given period is, i.e. how large the jitter is.
In addition, buffer levels during data exchange can be analyzed mathematically. This is particularly important for event chains. Furthermore, capacity utilization and end-to-end latencies should also be mentioned. In the latter case, it is very interesting to look beyond ECU boundaries—especially if it can be shown that an event chain can break off. For each question there is a suitable analysis method.

Peter Gliwa: What about the dependency on the operating system, on the scheduling procedure?

Dr. Ralf Münzenberger: The analysis itself is basically the same for all of them. Let's take priority-based scheduling as an example. The interrupt or task with the highest priority demands a specific amount of computing time and then passes on 'how much time is leftover' to the task of next lowest priority. This results in a cascade of calculations. Of course, it must also be taken into account whether an interruption is not longer possible at certain points in time, perhaps because the most recently executed task has disabled the interrupts.
Which interruptions are possible and what they look like depends on the scheduling policy of the operating system. Corresponding adjustments of the scheduling analysis then allow, for example, EDF or TDMA to be analyzed.

Peter Gliwa: That is, Earliest Deadline First (EDF), a method in which the task whose deadline would expire next is scheduled next, or Time-Division Multiple Access (TDMA), a time-slicing approach.
So far we've talked about scheduling, but earlier you mentioned data exchange buffer levels and end-to-end latencies. This means that data flows can also be modeled. How can I perceive that? Are the times for writing and reading defined relative to the start of the affected task or interrupt?

Dr. Ralf Münzenberger: Again, there are several approaches. One is the classic IPO approach: input, processing, output. At the beginning of a task, interrupt, or a runnable, the system reads, during execution it processes, and at the end it writes. This IPO approach is widely used in AUTOSAR CP applications. In ADAS systems . . .

Peter Gliwa: . . . that is, Advanced Driver Assistance Systems, or driver assistance systems. . .

Dr. Ralf Münzenberger: . . . and likewise with the systems in the realm of autonomous driving, we often observe that customers specify the data access times relative to the start of another event. So, for example, "read access occurs earliest after 1.2 ms and latest 1.9 ms after the start of thread XYZ".

Peter Gliwa: So, we've talked about how it works and how to use it. When do I, as a user, use static scheduling analysis as opposed to, for example, scheduling simulation?

Dr. Ralf Münzenberger: Because the proof is based on a mathematical approach, it automatically considers all theoretically possible cases. This is called a worst case analysis. This enables the user to answer the question "Are all my time requirements met?" with great certainty.
Static scheduling analysis is often applied to safety-relevant systems such as braking control or electrical steering systems. It is also used across several ECUs to investigate whether there may be unfavorable shifts in communication because clocks are not synchronized, or because of scheduling effects on communication buses.

This becomes especially interesting if there is a mixture of time-based scheduling and event-based scheduling. So, in order to take all possible situations into account, static scheduling analysis is used.

Peter Gliwa: The topic has been discussed in the academic environment for over 50 years and is very well researched. How is it that it is not yet widely present in the everyday life of the embedded software developer?

Dr. Ralf Münzenberger: First of all, there is a level of training deficiency. The subject is not an integral part of computer science and certainly not of engineering courses.
Another reason is that, for many developers, Microsoft Excel is still the number one development tool.

Peter Gliwa: (laughs)

Dr. Ralf Münzenberger: Yes, for many kinds of things as well as for scheduling analysis. Let me give you a concrete example. We were once called in to a customer whose system already had a considerable level of complexity. At some point, the customer opened an Excel spreadsheet with which he had determined the timing. The thing was full of macros and there was nobody in the room who still understood how it worked. But, in the end, it generated a number that they then used it to determine their architecture.

Peter Gliwa: Fascinating. Or rather, terrifying!
That means they programmed a scheduling analyzer in Visual Basic for Applications?

Dr. Ralf Münzenberger: Yes, indeed. There are different procedures that are sometimes simpler, sometimes more complex and, as a result, perform more or less well. There are, for example, those that exclusively support rate monotonic scheduling, i.e. periodic tasks without any spontaneous events such as interrupts. There the scheduling analysis algorithm is not very complicated. But there are hardly any embedded systems without interrupts...

Peter Gliwa: Another reasons why static scheduling analysis is still used so seldomly?

Dr. Ralf Münzenberger: A common misunderstanding is that too little information about the system is available in an early project phase. We hear feedback such as "I don't even know the execution times of my tasks yet."

Peter Gliwa: One of the strengths of static scheduling analysis is that I can set up my system and work with assumptions for the runtimes of tasks and interrupts. If the analysis for the system shows that all timing requirements are met, I can interpret the runtimes used as run-time budgets. If these budgets are adhered to during the later implementation of the tasks and interrupts, I know that my system is clean or safe in terms of timing.

Dr. Ralf Münzenberger: Correct. And of course the timing model will be adjusted as the project progresses. We have shown in countless projects, as well as in papers published together with customers, that it is better to take care of timing early on as it saves a lot of money and time. The term 'front-loading' is apparently still used in a negative sense. But the fact is, I have never seen a project where timing was tackled too early.

A further issue is that, if the issue of timing is not systematically addressed, the timing problems are then only detected very late in the project process. Additionally, they are often confused with functional errors. People think that the data is wrong, but it is only calculated at the wrong time or inconsistent. It is not uncommon for all kinds of screws to be turned until it somehow seems to work again. The fact that a new problem has been created at another place is only noticed *even later* in the project and then searching starts all over again.

Peter Gliwa: I would like to go deeper into that. Frankly, I'm surprised that you bring static scheduling analysis into play for the use case 'timing problems at advanced project stages'. My understanding is that, in such cases, the user must first *understand* what is happening and that the tracing is the tool of choice for this. I always say "I cannot model or simulate a problem I have not understood." How does that fit together?

Dr. Ralf Münzenberger: What we sometimes see is that the interaction of several factors is causal. Looking at a single trace can reveal a specific situation, but it does not allow us to see the dependencies.

The question is, can I isolate this locally or does this affect the overall design of the system? Scheduling analysis helps immensely here, especially since I can immediately see the effect of changes. Therefore, I can quickly evaluate possible solutions. If I simply turn a screw, I might solve the current problem, but may cause three other runtime violations. Thus, the supposed solution turns out to be none at all as I still have to keep looking.

Peter Gliwa: I understand the second part of your answer and I completely agree with you there. When it comes to finding the cause, I have usually had different experiences. Just as you say, people often don't realize that they have a timing problem at all. A specific functionality behaves differently than expected, or a calculation does not deliver the desired results. Often it takes weeks to discover that the algorithm itself is fine but the data changes 'out of the blue', or data is received too early or late. If I have a trace in which I display the received data and the runtime of the algorithm, in addition to the tasks and interrupts, I can see immediately what is going on. But without tracing, this is like looking for a needle in a haystack with your eyes closed.

If I then have an idea for how to resolve the problem, I can evaluate or validate the idea perfectly using static scheduling analysis. So I'm fully with you again.

Dr. Ralf Münzenberger: Let me explain this using a concrete project. The customer concerned had introduced a new time-controlled bus and had two control units connected to it. Suddenly he had repeated error-memory entries that messages had been lost. He had traces available for both the ECUs and the bus, and each trace showed that the timing was in accordance with the specification.

It was only the interaction of the non-synchronized clocks in a model—which, by the way, was created very quickly—that showed what was going on. Looking at the traces alone, he did not understand that the drift of the clocks plus jitter led to these errors.

Let me put it in a nutshell: measure/trace in detail, then analyze at the big picture level.

Peter Gliwa: Interesting. We were able to clarify the same situation several times in the past using traces. Obviously it is important to present the relevant information in one and the same trace.

Dr. Ralf Münzenberger: In the aforementioned project, we were then able to carry out a transient analysis, i.e. analyze the transitions, by targeted detuning of the clocks. This showed how often the problem can occur. From this they derived their $\binom{n}{q}$ validation and made the decision as to whether the vehicle could be released or not.

That's one thing you can actually only do with static analysis: specifically examine the error case and make statements about probabilities. At best, tracing can only do this with huge amounts of trace data.

By the way, we saw this same cause, in principle, once more in an ADAS project.

Peter Gliwa: Finally any recommendations or practical tips?

Dr. Ralf Münzenberger: Timing metrics belong on the dashboard of every project manager!

Peter Gliwa: Thank you!

As co-founder and managing director of INCHRON AG, Dr.-Ing. Ralf Münzenberger and his team have advised customers in more than 180 projects in the field of design, analysis, optimization, and testing of embedded systems with a focus on timing and performance. He dealt with these topics intensively during his doctorate at the University of Erlangen-Nuremberg. His interests focus on the areas of integrated design methodology across all relevant architectural levels (logical architec-  ture, system and software architecture) and the automation of tool chains.

Since 2003, INCHRON AG has successfully offered solutions that meet the high quality requirements of, and strict project schedules during, the development of embedded systems despite the exponentially increasing complexity. The INCHRON Tool-Suite supports worst case analysis and statistical simulation, as well as trace analysis and visualization. The main application areas are the design of robust architectures, virtual verification, and the verification of timing requirements during the test phase of single-core, multi-core, and distributed systems.

5.10   Optimization Using Evolutionary Algorithms

Although the method described in this section is not a timing analysis technique per se, it is often used by timing analysis tools to automatically optimize a system. Even for embedded systems with only a few tasks and interrupts, the scheduling can be so complex that timing parameters, such as the response time RT of a task, cannot easily be calculated. This makes optimization difficult, so finding a system configuration that minimizes the RT is not always easy.

Evolutionary algorithms are suitable for solving this type of problem and their mode of operation will be explained here.

Initially, the optimization target is specified, such as the minimization of the response time of a task as mentioned above.

The next step is to define the degrees of freedom, i.e. the parameters that may be changed during the course of the optimization. These could include the offsets of periodic tasks or the priorities of certain tasks.

Now the actual optimization starts. Simply put, the parameters forming the degrees of freedom are randomly changed, then an analysis is performed, and the resultant impact on the optimization goal is considered. Modifications to those parameters that serve to converge with the optimization goal are followed-up and the process starts all over again. Random modification of the parameters is similar to mutation in evolution. Successful modifications to the 'genetic makeup' prevail and, over several generations, the configuration improves and gets closer and closer to the optimization goal. If the optimization goal is sufficiently well implemented, or if a previously defined time is exceeded, evolution is stopped. Figure 63 shows this process.

The solution found is almost certainly not the optimal solution, but the optimal solution cannot usually be determined in finite time.

Often, several solutions are found that fulfill the individual optimization goals to varying degrees. Figure 64 displays a spider chart (it is strongly reminiscent of a spider's web, hence the name) that allows the user to visually compare the three different solutions in this 'case' with one another and with the initial state. In addition, the lines "Min" and "Max" mark the minimum and maximum values achieved in the solution set. Each vector that starts from the center represents an optimization goal. The smaller the distance to the center, the more optimally the goal has been achieved.

It stands to reason that, the more often the loop shown in Figure 63 is run, i.e. the more generations that are included, the better the result will be. The duration of the optimization thus depends crucially on the duration of the analysis. In scheduling simulation, the user typically faces a tightrope walk: if the duration of the individual simulation is too short, critical corner cases may not be captured. If they are too long, the duration of the optimization can be without end.

With static scheduling analysis this dilemma does not exist. The duration of the individual analysis depends solely on how skillfully it is implemented by the tool provider.

Figure 63 Flowchart of optimization using evolutionary algorithms

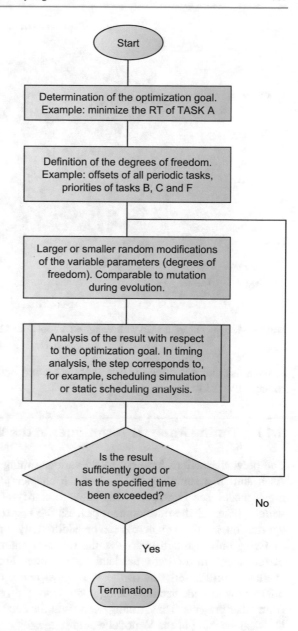

In practice, users are often offered several solutions and can then decide for themselves which solution they prefer. It is not uncommon for the second-best solution found to deliver only a slightly worse optimization result, but with very few changes to the available parameters that form the available degrees of freedom. If a system that is already running is optimized, every change to the configuration

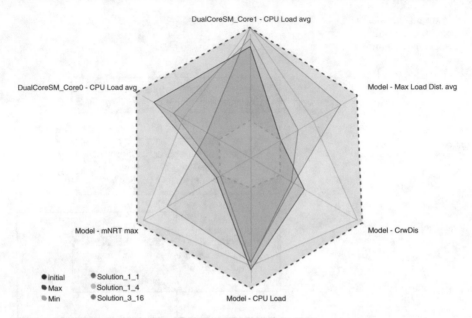

Figure 64 Spider chart created with Vector's TA Tool Suite [26]

poses a risk and it is often the case that the second-best solution is good enough if it reduces the risks.

5.11   Timing Analysis Techniques in the V-Model

The previous sections have described various timing analysis techniques in detail. As a sort of a summary, the end of the chapter places these in the context of the V-model, see Figure 65. This helps to clarify at which points in time and at which phases of the development process the respective technique finds its main application, as well as where it can be additionally used.

Being only a rough overview, the diagram naturally results in a high-degree of blurriness. In a strictly academic sense, there are no "initial portions of code available" on the left side of the V. In most real projects, however, the reality is different with code sections being taken from the previous generation designs or from other projects. Thus, timing analysis at the code level can start earlier than the classical teaching of the V-model would suggest.

Static Code Analysis   A static code analysis, in the sense of a WCET investigation that is based on an executable, requires completed and linked code. However, useful statements about the runtime can already be developed if the function to be examined is linked against a test application (with unit tests for example). To create the executable, the corresponding compiler toolchain must be available. For static code analysis, the analysis tool must support the processor used.

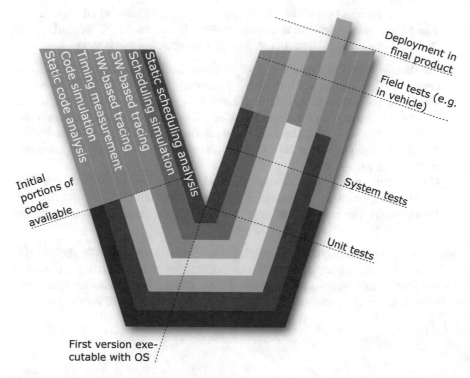

Figure 65 Use of the various timing analysis techniques in relation to the V-Model

Code Simulation   What applies to static code analysis also applies to code simulation. In addition, it is also possible to look at levels higher than the function level.

Runtime Measurement   As soon as an evaluation board with the desired processor and the corresponding compiler toolchain are available, runtime measurements can be performed on a PIL (**P**rocessor **I**n the **L**oop). This can be used during the further course of development as well as in the final product for runtime monitoring in regular operation.

Of course, pin toggling as a measuring technique is excluded from these considerations.

Hardware based tracing   As with runtime measurements, hardware based tracing also requires the availability of a running processor. Here, extensive analyses are possible both at the code level and at the scheduling level. Where the situation permits it, these analyses can also be extended to the HIL (**H**ardware **I**n the **L**oop), or even to the final product environment.

Tracing based on instrumentation of the software   Strictly speaking, tracing could be started at the same time as runtime measurements, but scheduling is usually an essential aspect of the analysis. Thus, the availability of a system on which the operating system is already running is a prerequisite. Similarly, trace data could be collected from use at the end customer, but this is unusual.

Scheduling Simulation   The focus of analysis using scheduling simulation is clearly the left side of the V-model. The analysis is largely independent of the availability of a processor, compiler, hardware, or software. Just a rough idea of what the system may ultimately look like is enough to simulate, analyze, and optimize it at a high level. The increase in information that results over the further course of development can be added to the model, thereby making it more and more detailed over time.

On the right side of the V, scheduling simulation can be used to consider corner cases, i.e. rarely occurring marginal cases that the tests have not covered. However, this use case pales in importance when compared to its use on the left side of the V.

Static Scheduling Analysis   Everything listed for scheduling simulation also applies to static scheduling analysis. Only the focus of the analysis is different as the worst case scenarios are considered more clearly. Static scheduling analysis cannot, however, provide an analysis of the normal behavior of the system.

Practical Examples of Timing Problems

6

Up until this point in the book we have discussed many basics and built upon them. The following sections focus on the practical aspects of timing analysis, each providing an example of a timing issue in a real project. It quickly becomes clear how varied the causes of timing problems can be, and also how differently the problems make themselves apparent. Unfortunately, there is no generally valid and detailed 'Guideline for solving timing problems dependent on their symptoms'. Here, only knowledge and experience can help—so perhaps the knowledge imparted through the following practical examples can serve as such a source.

6.1 Where Do All the Interrupts Come From?

A characteristic project progression was observed in a body segment automotive project that was developing a lighting control unit. This new generation of lighting control units was based on a previous generation. Many things had changed, such as the basic software, of which an updated version was being used. In addition, some new functionality had been added.

Only very rudimentary steps had been taken to validate the timing. A counter in the background task was in use to estimate the total system load.

It had become clear to the project manager that the system would be overloaded with the implementation of the functions that were still yet to be integrated. In order to first attain a better insight into the runtime behavior and optimize the timing, our team of specialists was commissioned to extend the software to include tracing using T1. The existing software was instrumented and the view of the very first trace downloaded from the ECU (see Figure 66) caused astonishment to the integrator. One of the interrupts was expected to occur once every 10 ms. Instead, a much more frequent occurrence of this interrupt was observed, as the figure shows. It shows a trace of about 7 ms in which, according to the project manager's presentation,

© The Author(s), under exclusive license to Springer Nature Switzerland AG 2021
P. Gliwa, *Embedded Software Timing*,
https://doi.org/10.1007/978-3-030-64144-3_6

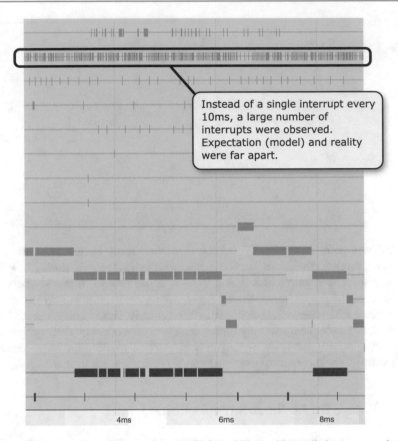

Figure 66 Trace section of approx. 7 ms in which the ISR should actually have occurred at most once

the ISR should have occurred exactly once or not at all. However, well over 200 instances of the ISR can be seen.

Initially, the inserted tracing feature was called into question but, after a short analysis, it was determined that the interrupt was mistakenly configured to trigger at a certain *pin state* and not—as actually intended—at a certain *change of state*, an edge. The ISR, i.e. the code that was executed as a result of the interrupt, was implemented in such a way that multiple executions did not cause a functional problem and the software operated without issue. However, the effect of this misconfiguration on the runtime requirements of the ISR was of course striking.

The solution was very simple and implemented within 10 min. After the interrupt was configured to trigger only on a rising edge, the subsequent traces always showed exactly one interrupt every 10 ms—as intended. The integrator did not dare to dream that over 10% CPU load would be recovered in such a short space of time.

The decisive lesson from history is that nothing can replace the view of the real system. Even if the project had previously taken a closer look at the timing, such as by means of scheduling simulation or static scheduling analysis, the interrupt in question would have been modeled to trigger every 10 ms. The results of such analysis would have given the project manager a false sense of security.

On several occasions, a simple look at the very first (scheduling) trace of a system has revealed problems and/or optimization potential that no one was aware of before.

6.2 OSEK ECC: Rarely the Best Choice

The series project described below was a little more dramatic. We were called in when sporadic functional problems and unstable communication was determined at an advanced stage of the project. The question of whether the system was possibly overloaded was answered as follows: "We implemented the ErrorHook in such a way that a reset is triggered as soon as it is executed. We have not observed any resets, therefore the ErrorHook is not being executed; therefore we do not have task overflows; therefore the system is not overloaded."

The term 'task overflow' is commonly used to describe a failed task activation. This occurs when the maximum number of possible activations for a task has been reached and a further activation attempt is then made.

Most systems do not provide for multiple activations in their configuration. This is good because there are a limited number of practical use cases for multiple activations. In most cases, the aim of allowing multiple activations is to cushion an overload situation that has already occurred, rather than to eliminate the causes of the overload. Therefore, the 'feature' multiple activations should never have found its way into the OSEK standard.

But, back to this mass-production project whose configuration excluded multiple activations. If a task is running in such a configuration, or if it has already been activated, the attempt to reactivate it directly triggers an error and the ErrorHook is called. The observation in this project that the ErrorHook was not executed was absolutely correct. However, the conclusion that there was no overload situation was not correct. How could this be?

To get to the bottom of this, you first have to look at the RTE and the operating system configuration. The **R**un-**T**ime **E**nvironment (RTE) will be discussed in more detail in Section 10.1.3. Listing 24 shows an example of how most AUTOSAR RTE code generators handle periodic runnables by default. Instead of generating a separate periodic task for each period that occurs in the configuration, all runnables are loaded into a single task. This task is configured as an ECC task (Extended Conformance Class, see Section 3.2 on page 40). It contains an endless loop (for(;;)), which makes it a non terminating task.

Listing 24 Example RTE task as it is often generated

```
1  TASK(Task_B)
2  {
3    EventMaskType ev;
4    for(;;) // non-terminating ECC task
5    {
```

```
6      (void)WaitEvent(      Rte_Ev_Task_B_10ms |
7                            Rte_Ev_Task_B_5ms );
8
9      (void)GetEvent(Task_B, &ev);
10
11     (void)ClearEvent(ev & ( Rte_Ev_Task_B_10ms |
12                             Rte_Ev_Task_B_5ms ));
13
14     (void)Com_ReceiveSignal(TempS2_Rx, (&TempS2_Tx_local));
15     if ((ev & Rte_Ev_Task_B_10ms) != (EventMaskType)0)
16     {
17       CanNm_MainFunction(); // Runnable
18       CanSM_MainFunction(); // Runnable
19       my10ms_worker_runnable(); // Runnable
20     }
21
22     if ((ev & Rte_Ev_Task_B_5ms) != (EventMaskType)0)
23     {
24       CanTp_MainFunction(); // Runnable
25       CanXcp_MainFunction(); // Runnable
26       my5ms_worker_runnable(); // Runnable
27     }
28     (void)Com_SendSignal(Torque_Tx, (&Torque_Tx_local));
29   }
30 }
```

At the beginning of each loop run the task is put into its Waiting state by the function call WaitEvent(...), i.e. it is put to sleep. The execution of the task is only continued when at least one of the two events that are being waited for (Rte_Ev_Task_B_10ms and Rte_Ev_Task_B_5ms) is 'set' outside of the task. In the further course of the loop body, the runnables now execute depending on *which* event is set. As per their naming, all runnables belonging to the event Rte_Ev_Task_B_5ms were called every 5 ms and all runnables belonging to the event (Rte_Ev_Task_B_10ms) were called every 10 ms.

So far so good. But what happens if the execution of the runnables takes longer than expected? The reasons for this can be very different. Perhaps some of the runnables have an unexpectedly high core execution time, or the task was interrupted by other, higher-priority tasks or by interrupts.

Let's assume a single loop pass takes 16 ms due to a combination of these reasons. During this time the 5 ms event is set two more times, and the 10 ms event is set one more time. However, the executions of the runnables that are actually linked to these events does not take place and they are simply lost. This occurs without it being regarded as an error by the operating system. Setting a new event that has already been set *does not* cause the execution of ErrorHook, unlike the reactivation of an already activated or running task would.

In the end, this series project did not have a loop body with an execution time of 16 ms but one of 26 ms, and several ECC tasks were involved. Figure 67 shows the resultant trace that has been manually edited with red markers. These markers indicate when events are set again without the associated runnables being run. In a configuration with terminating tasks that are activated each time they are to be executed, the red markers would indicate failed task activations together with a call to `ErrorHook`.

The discovery initially caused a shiver in the customer's spine. The recognition of not having 'task overflows' in the software was replaced by the certainty of being exposed to massive runtime problems. However, the trace shown in Figure 67 immediately revealed the main cause: the task responsible for communication had an exorbitantly high CET whenever the service discovery was running. The task can be seen pretty much halfway down the trace and has many, very long dark green running blocks.

Now the task could be rebuilt and optimized at the code level, whereby the functional problems were solved and the communication became stable again. The recommendation to convert the non-terminating ECC tasks into (normal terminating) BCC1 tasks could not be implemented as the project was already too far advanced for a significant change of this kind.

The monitoring of timing parameters was introduced to ensure that bottlenecks due to unexpectedly high CETs could immediately be detected over the remaining course of the project. Exceeding the specified CET or CPU load now results in an error buffer entry together with a snapshot trace that visualizes the critical situation precisely, along with the timing violation.

6.3   Rare Crashes 17 min After Reset

If you ask an experienced firefighter about the nature of his job, you will be provided a wide range of situations, from the cat stuck in the tree to the big fire in a factory—there's bound to be a lot of stories to offer.

The project described here was a major fire in the figurative sense. The start of series production for a vehicle was endangered, the management through to the chairman of the board of the car manufacturer was involved, and a task force, which had meanwhile grown to 18 people, had been trying for more than six months to master the problem.

When I was called in and the slides were presented, I noticed how routinely this presentation was prepared. Obviously it had been presented many times before. On slide five of about twenty it was described in more detail how the problem became apparent. In the past two years it had appeared a total of six(!) times. As a result, the operating system seemed to 'freeze' and refused to continue executing tasks. Interrupts, however, continued to occur.

Almost in a subordinate clause a remark was made that, in five of the six cases, the problem occurred after about 17 min, while it had once occurred after about 34 min. I was electrified by this and did not want to wait for the end of the presentation. Several times 17 min or a multiple of 17 min—this could not be a coincidence. What was special about this time span for the given control unit and its software?

A look at the configuration of the clock source and various timers revealed a connection. Some of the timers had a tick duration of 237 ns and a width of 32 bits.

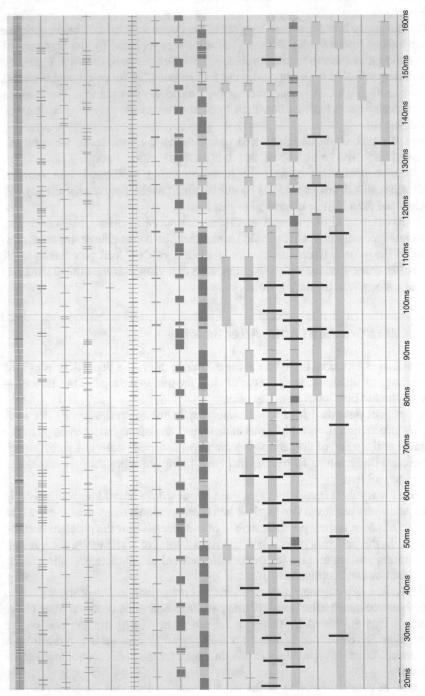

Figure 67 Unnoticed lost runnable executions due to non-terminating ECC tasks

That meant the timer value was counted up or down every 237 ns and, after 2^{32} counting steps, an overflow or underflow occurred. And, with a bit of math: $2^{32} \cdot 237$ ns ≈ 1018 s ≈ 17 min. I was sure I was on the right track.

I called in my colleague Dr. Nicholas Merriam (Nick), a seasoned embedded software expert. He is very familiar with the PowerPC architectures. Soon, the decrementer interrupt came into view as the cause. The PowerPC architecture was never designed to be an embedded processor. At the time, Motorola had developed it for desktops and servers and later wanted, short term, to gain market share in the embedded sector. Thus, some 68000 peripherals and some interrupt logic were taken out of the drawer and connected together. The microcontroller for the embedded arena was ready—this was long before Infineon had its TriCore ready for the market. PowerPCs were initially produced for the trash can for months because, initially, Motorola could not get the production process for the on-chip flash under control. However, when it was finally available, there was little serious 32-bit competition on the market.

But, back to the real problem. The core of the PowerPC itself has only two interrupt sources: the external interrupt, to which the aforementioned interrupt controller is connected (and which enables further interrupt sources), and the decrementer interrupt. The decrementer is a 32-bit timer that simply counts down after a reset without any further configuration and triggers an interrupt when changing from 0x00000000 to 0xFFFFFFFF. So, this interrupt was a good candidate for the cause of the problem.

Since, as mentioned above, the problem could only be observed six times during the course of the project, we considered how we could reproduce it more regularly. In general, this consideration should always be made when dealing with rarely occurring problems. There is not always a solution, but in our case it was quite simple. The crucial idea Nick and I came up with at lunch, after we had spent a few hours of the morning on the in-depth analysis, was to set the timer value to a low value. This would ensure that the interrupt was triggered many times per second instead of only once every 17 min. This allowed the problem to be reliably observed within a few minutes after reset.

Further analysis showed that, in the case of an error, the data of the operating system became inconsistent and, finally, a review of the interrupt service routine (ISR) of the decrementer interrupt revealed that it had been forgotten to save one of the registers on the stack. The ISR had been programmed manually in assembler and the error had not been noticed in the previous reviews.

No sooner were *all* registers in the ISR saved correctly, the problem no longer occurred—even if the ISR was executed several times per second due to the described manipulation of the timer.

The SOP (Start of Production) was saved and what remained was the realization that, every now and then, the entire armada of analysis tools is useless. Sometimes (or even often?) only the correct mix of experience, inventiveness, and the right person to talk to over a good lunch can help.

6.4   Missing or Duplicated Sensor Data

The problem described in this section can safely be called one of the classics among timing problems. In fact, it is a classic in two respects because, firstly, it has been seen repeatedly in different projects for decades and, secondly, because it has the typical characteristics of a timing problem. The first of these is that it occurs sporadically. The second is that, in most cases, a functional problem is initially thought to be the issue. And, thirdly, it is very easy to prevent if you only consider the issue of timing from the start of the design.

An example of this is a project from the chassis division. The OEM observed at a subsystem test station that CAN messages were being lost from time to time. What seemed strange initially was that the dropouts were observed quite regularly every few minutes. Even more curious was that the time interval between the dropouts was different for different ECUs. On one unit the problem occurred pretty much every 14 min, on another one it was every 18.

By means of tracing the problem could be found quite quickly. The CAN reception routine was instrumented in such a way that a 'user event' was entered into the trace for each reception of the message concerned. This is displayed as a vertical line in the trace and the time of receipt can be related to the scheduling, i.e. the execution of tasks, interrupts, and runnables.

The traces indicated that the affected message was received every 10 ms—with some jitter. This means that they sometimes came a little earlier, but also sometimes a little later than 10 ms. The data contained in the message was used by a periodically executed runnable for calculations. Again, not surprisingly, the execution was subject to jitter. The trace showed that, in the case of an error, the calculation took place, the message was received shortly after that, the message was received again around 10 ms later, and the resultant calculation used the newest value with the previously received data never being used. Here were the 'lost' data packets.

In another error case it was observed, as well as being clearly visible in the trace, that data was being used twice. Data was received, a calculation was performed and, before the next data point was received, the calculation ran again using the data that had already been processed. Figure 68 shows both error situations in one image.

So, how is it possible that reception and calculation do not always mesh like gears, although both occur with a period of 10 ms?

The received message was previously sent by another participant on the bus. This participant is itself a control device, or at least a sensor with its own logic. Both the transmitter and receiver have their own crystal which are used to generate the processors' clocks. The 10 ms period with which the message is transmitted therefore had a different timebase to the 10 ms period of the system which was receiving the message. Crystals have manufacturing tolerances, so they differ from their nominal frequency. This deviation also depends on environmental factors, such as temperature. The two relevant periods in the project will certainly not be exactly 10 ms, but one will be perhaps 10.000038 ms with the other lies at 9.99999925 ms.

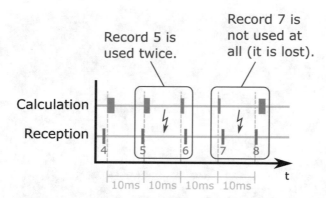

Figure 68 Visualization of duplicate reception and lost data

Over time the two timebases *drifted* past each other. Additionally, transmission and reception are subject to jitter. Only the bus's arbitration was capable of ensuring that a message would be delayed if a message with a higher priority was currently occupying the bus.

These two effects, drift and jitter, combine to cause the data loss and duplication as shown in Figure 68. By the way, the numbering of the messages shown in the figure had also been realized in the project. The message contained four bits that were used to implement a message counter. Before each message was sent the counter was incremented by one, making it easy for the recipient to check whether there was a loss of data or double use. In the traces, the counter was displayed below each user event, as shown in Figure 68.

In Listing 25 that follows, it shows the source code for a simple simulation of the combination of drift and jitter described above. It is only for illustration purposes but allows the reader to try the simulation for themselves and experiment with it.

Listing 25 Source code for the illustration/simulation of drift

```c
// simulate/show how two tasks with similar periods drift
#include <stdio.h>
#include <string.h>
#include <stdlib.h>
#include <unistd.h> // required for usleep

// one column is one character and represents one millisecond
#define NOF_COLS        (120)     // number of columns
#define NOF_ROWS        (2000)    // number of rows

#define PERIOD_1        (20.0f)   // period of calc in ms
#define PERIOD_2        (20.01f)  // period of RX in ms
#define INITIAL_OFFSET  (8.7f)    // related to RX
#define JITTER_FACTOR   (2.5f)    // related to RX

int main(void)
{
```

```
18   char buf[NOF_ROWS * NOF_COLS];
19   int i;
20   double pos, jitter, t = 0;
21
22   memset(buf, ' ', sizeof(char) * NOF_COLS * NOF_ROWS);
23
24   while (t < NOF_COLS * NOF_ROWS) { // firstly calc
25     buf[(int)t] = '|'; // a "|" represents calc
26     t += PERIOD_1;
27   }
28
29   pos = t = INITIAL_OFFSET;
30   while (pos < NOF_COLS * NOF_ROWS) { // now RX
31     if (buf[(int)pos] == ' ') {
32       buf[(int)pos] = '-'; // a "-" represents RX
33     } else {
34       buf[(int)pos] = '+'; // "+" indicates calc and RX
35     }
36
37     jitter = JITTER_FACTOR * ((double)rand()/
38         (double)RAND_MAX - 0.5f);
39
40     t += PERIOD_2;
41     pos = t + jitter; // only RX jitters
42   }
43
44   for (i=0; i<NOF_ROWS; i++) { // now print to stdout
45     fwrite(&buf[i*NOF_COLS], sizeof(char), NOF_COLS, stdout);
46     usleep(1000); // slow down output a bit
47     printf("\n");
48   }
49
50   return 0;
51 }
```

In the simulation two events, let's call them 'reception' (in code "RX") and 'calculation' (in code "calc") occur periodically. The calculation occurs every 20 ms and the reception on average every 20.01 ms. The reception is also subject to a jitter of 2.5 ms ($\pm$1.25 ms). The output is generated in text-form in a console, where each character position represents 1 ms. At the end of a line a line break is simply inserted. All macros ("#define") can be customized. The reception is displayed with a horizontal line (-) and the calculation with a vertical line (|). If both events roughly coincide, a plus (+) is displayed. If an integer multiple of PERIOD_1 is selected for the number of characters per line NOF_COLS, the calculation times in the different lines occur one below the other and the drift in PERIOD_2 is easier to recognize (see also Figure 69). In the online support accompanying the book you can also find a video showing the execution of the simulation.

In Figure 69 the offset of the two periodic events is so large that they always occur alternately despite the jitter. A short time later this is no longer the case and then it looks like in Figure 70. An alternating sequence is not always given and data loss or double received data will occur.

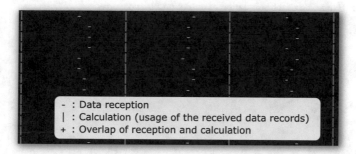

Figure 69 Simulation output with reception and calculation interleaved

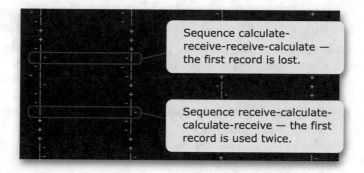

Figure 70 Simulation output with the reception and calculation overlapping

The short periods in which the calculation 'overtakes' the reception, and in which the problems occur, are repeated cyclically. The behavior is similar to a beat in acoustics as can be heard, for example, when two strings of a musical instrument are slightly detuned with respect to one another. Here, too, we are dealing with two frequencies that are close together. The two tones overlap and cyclically amplify or attenuate each other. This is perceived as beating, a tone whose volume increases and decreases at a low frequency.

If you extend the code from Listing 25 before `return` with the lines shown in Listing 26, you obtain a simple analysis of the simulation. All times at which a problem occurred are then displayed. It shows that every 40 s a batch of duplicate received or lost data is observed.

As with beating, the difference in frequencies plays a role here. With the values from the simulation example we obtain the following results: if $T_1 = 20$ ms and $T_2 = 20.01$ ms, $f_1 = 50$ Hz and $f_2 = 49.97501249$ Hz. The frequency f_S of the 'beat' is calculated as follows. $f_S = |f_1 - f_2| = 0.024987506$ Hz and thus the problem occurs on average every $T_S = \frac{1}{f_S} = 40.02$ s.

Listing 26 Extension of the simulation to output problem occurrences

```
1    char last = ' ';
2    for (i=0; i<NOF_ROWS * NOF_COLS; i++) {
3      if (buf[i] == '-') {
4        if (last == '-') {
5          printf("double - at %dms\n", i);
6        }
7        last = '-';
8      }
9
10     if (buf[i] == '|') {
11       if (last == '|') {
12         printf("double | at %dms\n", i);
13       }
14       last = '|';
15     }
16   }
```

The much longer period duration of the beat mentioned at the beginning of the series project is due to the high accuracy of the crystals used. A common crystal as clock generator delivers the desired frequency with a very small error of about 30 ppm. The 'ppm' stands for parts per million (10^{-6}). For the simulation example, T_2 should deviate by only 0.0006 ms instead of 0.01 ms. This period of beat results in a $T_S = 27.78$ min if T_1 corresponds exactly to the nominal period. In practice, however, this value will also deviate so that even with maximally detuned crystals, the errors occur at the earliest every 13.89 min.

The more accurate the crystals are, the greater the distance between which groups of faults will be observed.

The problem and its cause are thus described in detail. So, what are the possible solutions? There are two approaches to solving the problem:

Synchronization   Since the crystals cannot be synchronized, the synchronization must be implemented in the software. The following list, which is certainly not complete, shows different approaches to synchronization.

- If both transmitter and receiver use AUTOSAR, they can be synchronized via the "Synchronized Time-Base Manager" [24]. This approach is quite a major intervention in the architecture of the overall system. As a result, the operating systems of the ECUs involved are synchronized— including the cyclically executed code components responsible for sending, receiving, and calculating.
- Instead of attaching the calculation to a cyclical task, and thus coupling it to the crystal of the receiving ECU, the calculation could be triggered each time data is received. This approach may, however, only shift the problem. Assume that the results of the calculation need to be used by other tasks in the receiving ECU. If these tasks are also cyclical tasks, the problem of beating will occur between the calculating task and the processing task.

Independent of this aspect, it would be advisable to check the correct reception of the data with the 'calculation upon receipt' approach in order to detect, for example, the loss of data.

- Instead of, as just described, synchronizing the receiver to the transmitter, the transmitter can also be synchronized to the receiver. The receiving ECU could explicitly request the data from the receiver before each calculation.

  A combination of these approaches is also conceivable and useful. The receiving ECU requests data cyclically and triggers the calculation upon receipt of the data. The task that continues processing is again cyclic and first checks whether exactly one calculation has been performed since the last request.

Oversampling   If each measurement were to be sent twice, with doubly-received measurements being discarded, no data would be lost. Obviously, the load on the network increases with this approach and the algorithm needs to be modified accordingly.

6.5   In a Race with the Handbrake On

From some missions you emerge as a rescuer in times of need, while from others you only contributed to the success of the project to a small extent. Other times it turns out that you only witnessed the rescue of a project. This is what happened in the mission that will be briefly described here.

On a sunny morning in September 2019, a call for help came from a customer who had been using our T1 measurement and trace solution for several months. His control unit software was no longer stable and traces could no longer be downloaded. A general overload of the processor was already identified as the cause.

By chance, I was able to drive to the customer on the same day. Just as I was about to start the tests, one of the developers came to the test site with a new software version. In this version the program cache (P-Cache) had been activated for the first time. I couldn't believe my ears—had the P-Cache not been used before?

The processor used was a second-generation Infineon AURIX. The rather large project executed large amounts of code from the shared flash, and access to it was correspondingly slow. With the P-Cache enabled, the code was now executed about *four times faster* on average and, suddenly, not only the fetching of the traces worked again, but the functionality of the software was also restored.

For the last few months the project had been 'running with the handbrake on'.

After this experience, the question was raised if other similar measures could be taken. The D-Cache (data cache) accelerates data accesses in a similar way. I was not present when it was activated, but I emphasized that—unlike when using the P-cache—caution is required (see Section 2.5.3).

If you follow just one very simple rule in this context, as shown in the tip below, nothing can go wrong:

Hint *Data accessed by multiple cores must be located in a memory area for which*
the data cache is disabled. This may not always provide the best performance but,
as cache usage increases, it avoids having to implement cache consistency and
coherence by other means.

6.6   Measurement Delivers WCET Results Greater Than Those from Static Code Analysis

Within the framework of a research project, tool manufacturers of different timing
analysis techniques and two universities came together to create interfaces between
the tools and to work on a uniform methodology.

I was able to persuade two customers from the automotive sector to each provide
one of their projects for the research project as the subject for the timing analyses.
This allowed all analysis approaches to operate on real data, which made the
research project much more practice-oriented. Unfortunately, research projects work
far too often exclusively with academic example data or example projects. Then,
when applying the researched methods or algorithms in real life, a large number of
these new approaches fail.

For example, one of the participating universities had to painfully realize that
its approach to static code analysis is not practicable in the automotive sector. The
approach was based solely on the analysis of the source code that had to be available
in full for the analysis. However, hardly any automotive project has a project partner
who has access to the *entire* source code of the project. Usually, different suppliers
and the automobile manufacturer work together and each protects its IP (Intellectual
Property) by only exchanging code at object code level (see also Section 9.2 and
especially Figure 99).

However, in this section the main focus is not the static code analysis of said
university. Of more interest was the following incident that was related to static
code analysis based on the executable.

In the research project our timing analysis approach, based on tracing and
measurement, brought us to the point where we wanted to compare our results with
those of other tools. We were very surprised when we came across measurements
of the net runtime (CET, **C**ore **E**xecution **T**ime) of a function that were *greater*
than the upper bounds calculated by project partners using static code analysis. In
other words, we measured values greater than the mathematically possible WCET
(**W**orst-**C**ase **E**xecution **T**ime).

At first, it was very difficult to discuss the situation with the developers of the
static code analysis used since their only reaction was that we were obviously
measuring incorrectly. So, we checked our measurements again and the timebases
used. On the one side there was the timer that was used to generate the timestamps.
On the other was the PLL configuration of the processor which came under the
microscope. Perhaps the whole software was running at a different speed to that
expected?

In order to check this second aspect very simply and reliably, there is an almost trivial procedure. One provides a function that is called cyclically—for example, the code of a periodic task that is activated every millisecond—with an integer counter. This variable is incremented by one each time the cyclical code is called ("variable++;"). If you now stop the processor with a debugger connected, set the variable to zero, then continue the execution of the software and stop again after exactly 1 min, the counter should now have a value of 60,000. A simple wristwatch is sufficient as a time reference for the 1 min. With this measurement setup the measurement error remains below 1%, if you take some care.

With the control measurement of this kind in the research project the error was negligible and the traces for the "1ms Task" also showed a period of almost exactly 1 ms. Thus, the whole chain of tracing from acquisition, storage, transmission, evaluation, through to visualization was correct.

Two weeks after the first inquiry, and after the presentation of the control measurements, the research partner's developers finally took the matter seriously and quickly realized that the hardware setup was incorrectly modeled. They had forgotten to specify the wait states for the flash from which the program code was executed. All previous results of the static code analysis were thus far too low, far too optimistic.

The project partner dismissed the whole thing with a shrug of the shoulders. And, yes, it was only a research project. However, this faulty configuration could also have occurred in a safety-relevant project, such as in the aviation sector. For sure, such situations happen from time to time and, as a result, timing problems go unnoticed onto the road, into medical equipment, airplanes, power plants, and so on.

From this experience one can deduce that any simulation, any model-based analysis, is only credible if the key data of these approaches are verified by means of observation or measurement on the real system. This verification need not be extensive, but it is essential to ensure that the model or simulation reflects reality sufficiently well in its core aspects.

A customer who, after a long period of troubleshooting and considerable additional costs, had to recognize a gaping gap between his model and the actual behavior of his software, once put it this way: "The difference between theory and practice is greater in practice than in theory."

6.7   Network Management Messages Appear too Soon

"Tell me, how do you actually send out your network management messages? When we receive your messages, we regularly get diagnostic trouble code (DTC) entries on our ECU," said one developer at lunch to his table neighbors who were developing software for another ECU. There came a general shrug of the shoulders with the response that this was probably a problem with his own software. However, a closer examination of the case showed that this assumption was wrong.

According to the specification, the aforementioned network management messages should be sent every 10 ms. A deviation of 1 ms more or less was permissible.

The receiving ECU checked this timing requirement and occasionally found that the time difference between two consecutive messages was less than 3 ms. The timing requirement was thus violated and a DTC entry into the error buffer was made.

Obviously, the sending control unit was responsible for the problem and the shrug of the shoulders mentioned at the beginning continued for some time by the responsible colleagues during the search for the cause.

The targeted use of the previously integrated software-based measurement and trace solution was finally able to shed light on the situation. The sending of the message was initiated by a specific runnable of the network management system. Once the delta time (DT) of this runnable was configured as a timing constraint in the runtime analysis tool, the monitoring mechanism triggered and provided a trace, in the middle of which the problem was now openly visible (see Figure 71). Whenever switching from one application mode to another, the cyclical sending of the message went 'out of step'. Each application mode had its own set of tasks and the time difference between the last call of the previous application mode and the first of the following one was too short—by a good 7 ms.

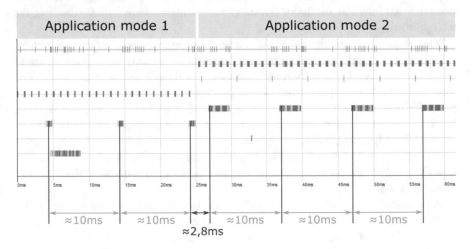

Figure 71   Timing violation when changing application mode

The solution to the problem was very simple. Only the offset of the cyclical task in the subsequent application mode had to be increased by 7 ms.

To ensure that the jitter remained within these specified limits in future, the automated timing tests were supplemented by the creation of dedicated 'point of interest' traces. All conceivable switches between application modes were executed in a targeted manner and the tracing was performed in such a way that the switches were captured in the downloaded traces. The verification was also automated and the traces were saved as part of the release documentation.

As the project progressed, the problem did not reoccur and the developers were able to move on to more pleasant topics over lunch.

6.8   Seamless Timing Process in a Mass-Production Project

Strictly speaking, the example given in this section is out of place in this chapter, because it does not describe a problem at all, but rather a success story.

In 2009 BMW requested quotes for an ECU project in the chassis domain and formulated the timing requirements using the then still relatively new standard "AUTOSAR Timing Extensions" (TIMEX [30]). TIMEX is described in more detail in the Section 10.3.

In addition to the formal description of the timing requirements, these requirements had to be imported directly into the measurement and trace technology and then verified automatically. In addition to the timing requirements regarding the parts of the software developed by the supplier, BMW also provided formal guarantees regarding the timing of the software provided by BMW—mostly minimum/maximum CETs for their runnables.

It was precisely this second aspect that resulted in the very formal approach being extremely well received by the supplier who was eventually awarded the contract. "At last an OEM is able to define what its share of code requires in terms of runtime," was the message from the project manager on the supplier's side. And indeed, over the course of the project the automated runtime measurements went 'red' a few times and it was BMW's turn to optimize their software components.

This was very helpful and contributed to the great success of the project. Despite high processor utilization, there were no delays in the project that could have been attributed in any way to the embedded software timing.

Shortly before the ECU went into mass production, the successful approach of a seamless tool chain with formal specification of timing requirements was presented at ERTS, the "Embedded Real Time Software Exhibition and Congress" in Toulouse in 2012. The corresponding publication [31] is available for download and can also be found in the accompanying online resources for the book.

6.9   Timing Analysis Saves OEM €12 m

Timing analysis does not yet seem to be sufficiently anchored in the courses of computer science, electrical engineering, or information technology. If every engineer were to start his or her professional career with sufficient knowledge in this field, the use of the correct timing analysis technique would be less dependent on chance.

In the project that will be the subject of this section, chance was a major factor in the fact that, in the end and thanks to timing analysis, resulted in one of the biggest cost savings for a single project.

A befriended software developer, with whom I had already solved timing problems in some projects, was called in to a project from the body division. His employer—a European car manufacturer—maintains good cooperation at the technical level with his suppliers and this ultimately paid off for the project described here.

The project was the last of a whole generation. Planning for the next generation had already begun. So far, the projects used a 16-bit controller, which was to be replaced by a 32-bit controller in the next generation. For the last project, another small functional boost was planned that, unfortunately, pushed the system to its limits. The task of the software developer from the automobile manufacturer was now to use his experience in the timing environment to support the supplier in bringing the project safely to fruition.

For the detailed analysis he brought GLIWA on board and our runtime measurement and trace technology T1 was quickly integrated. The bottlenecks caused by the additional functionality were also quickly eliminated.

Instead of wrapping the job up with the good feeling of having solved the task quickly, the developer wanted to develop a full understanding. The timing analysis was continued and more and more potential for optimization was revealed. The subsequent measures that were taken were so successful that more and more computing time was freed up while the CPU load sank further and further.

Finally, a point was reached where the developer informed his management that he considered it quite possible to run the functionality of the successor generation on the old 16-bit hardware. Further investigations followed, including tests with early versions of the software planned for the next generation.

The end of the story was that the next generation was no longer required and the entire range of functions intended for it could be handled with the existing hardware. The automobile manufacturer was able to save the development of an entire hardware generation, including the costs for the changeover, not to mention the higher production costs the 32-bit variant would have incurred.

Months later, a manager of said car manufacturer took me aside at a congress and told me that the whole action had saved his company about 12 million euros. If this is not motivation enough to give timing analysis a little more room at universities in the future, I don't know what is!

6.10   Summary

This chapter was a rather colorful compilation of field reports. What they all have in common is a reference to the topic of timing and the realization that a lot of development time, money, and trouble can be saved with appropriate skills in this area.

Multi-Core, Many-Core, and Multi-ECU Timing 7

It is common knowledge that it is impossible to shorten a pregnancy to 1 month by entrusting nine women with the task. Nevertheless, this is attempted again and again—figuratively speaking—in multi-core projects.

The effort required to switch from single-core to multi-core in *embedded* projects is almost always drastically underestimated. Yet, most of the processors we deal with in everyday life are multi-core processors. No smartphone, tablet, laptop, or desktop PC today has a single-core processor. One would think that the development of multi-core software is the rule, rather than the exception, and should therefore run smoothly.

The main reason why multi-core causes significant timing problems in most embedded projects is due to the different types of parallel execution. Parallelism can take place at the instruction level, at the function level, or at the application level. This chapter will cover the basics on multi-core and describe the different types of parallelism. It will also become clear why there is no tool that automatically converts single-core code to multi-core code. Although such 'C to C compilers' are available on the market and have been the subject of academic research for decades, they still have not proven that even moderately complex, single-core projects can be effectively migrated to multi-core.

Some of the aspects dealt with below are summarized on a DIN-A1 poster. This poster is available for download as a PDF [32]. It is also available together with the other online book material.

7.1 Multi-Core Basics

Since the advent of software there have been calls for more performance in hardware, in the processors. Over the years, embedded processors have had their clock speeds steadily increased while also becoming more powerful in other ways, such as through the introduction of caches and sophisticated pipelines. However, at

© The Author(s), under exclusive license to Springer Nature Switzerland AG 2021
P. Gliwa, *Embedded Software Timing*,
https://doi.org/10.1007/978-3-030-64144-3_7

some point it becomes difficult to simply make a processor 'even faster'. The simple increase of the processor clock alone results in a wide range of challenges.

For one thing, there is the power consumption. A processor has many millions of transistors and, with each switching operation, some current flows. Here, the transistor behaves similar to a capacitor that has to be charged. At the processor's top speed, $V \propto f$ shows that the voltage required is proportional to the clock frequency. If one simplifies all the switching transistors down to a single capacitance C, the following describes the power: $P \propto V^2 \cdot f \cdot C$ and, thus, $P \propto f^3$. The required power, therefore, increases by a power of three as the frequency is increased.

The power converted by a processor generates heat that has to be dissipated. At best, this can be challenging and expensive. In the worst case, it can lead to device failures. In the past, for example, insufficiently dimensioned heat dissipation of an instrument cluster (automotive instrumentation panel including speedometer) has led to parts of the display melting under certain conditions.

Another challenge at high frequencies is electromagnetic compatibility (EMC). High frequency and high power is precisely what is required to build a radio station. The electromagnetic waves that a highly sophisticated processor transmits in its electrical environment are the same and result in many problems. Close proximity signal paths, or even those within other electrical systems, are disturbed because their circuit paths act like antennas by coupling those signals. Such EMC problems are extremely difficult to analyze. Sometimes they occur very sporadically and only in individual examples of the same product.

So, how can the need for more computing power be satisfied? One possible way to obtain more computing power with a moderate increase in electrical power is to use multiple computing cores working in parallel. If these cores are placed on the same processor, on the same piece of silicon, we term it a multi-core processor.

7.1.1   Amdahl vs. Gustafson: Who Is Right?

The computer expert Gene Amdahl has found that the potential for increasing (software) performance largely depends on what proportion of that software has to be processed sequentially. This impacts all software, so even a doubling of the number of computing cores will, at some point, no longer bring any practical speed advantage. The increase in speed approaches an asymptote as the number of cores increases.

Opposing Amdahl's view is the statement of John L. Gustafson who says that programmers tend to take full advantage of the hardware provided to them.

If a developer gets a processor with twice as many cores, they will expand the software accordingly so that the performance of the additional cores are also exhausted. Accordingly, the increase in performance is linear to the number of cores.

So, who's right? Amdahl or Gustafson?

Embedded software projects typically have many elements that are difficult or impossible to parallelize. This is either because the functionality simply does not allow it, or because larger portions of code from previous single-core projects have to be accommodated. A control algorithm, a complex state machine, or a gateway that transmits messages between different communication buses, is not comparable to a rendering engine that can easily be distributed over hundreds of (graphics) computing cores.

Perhaps unlike the academic world, the task of an embedded system is clearly defined. Just because more powerful hardware is available, it does not result in a change of the definition of the task.

For these two reasons—large quantities of code that are difficult to parallelize and a clearly defined range of functionality—Amdahl's approach to embedded software fits much better.

7.1.2   CPU Cores: Homogeneous, Heterogeneous, or Lock-Step?

A processor is a multi-core processor if it contains several computing cores. However, classic single-core processors, such as the NXP PowerPC or the Infineon TriCore TC1767, have smaller cores implemented for specific tasks in addition to the main computing core. However, nobody would think to name these processors 'multi-core processors'. Thus, the above definition needs to be slightly modified: a processor is a multi-core processor when it has several *main computing cores*.

If a processor has several main computing cores, the question arises as to how the cores relate to each other. Are they identical in structure? Are they significantly different? Figure 72 shows some different multi-core approaches.

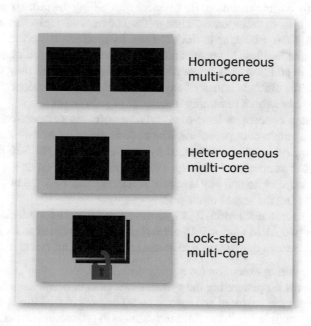

Figure 72 Different multi-core approaches

Homogeneous multi-core: Several cores of the same type. Examples:

- NXP 5xxx PowerPC
- Graphics cards with a large number of shader units
- Infineon AURIX TC277: this has two TC1.6P (performance) cores

Heterogeneous multi-core: Several (main) cores of different types. Examples:

- Renesas R-Car H3 (with Cortex-A57, Cortex-A53, Cortex-R7)
- Infineon AURIX TC277: in addition to two TC1.6P cores it has a TC1.6E (efficiency) core and several smaller cores.

Lock-step multi-core: The software is executed in parallel on two cores that are identical from a software point of view. This allows transient errors to be detected. Section 7.1.3 covers this topic in more detail.

- Texas Instruments TMS570
- Infineon AURIX TC277: one of the two TC1.6P cores and the TC1.6E (efficiency) core have a 'Checker Core' (Infineon term for its lock-step cores).

7.1.3   Lock-Step Multi-Core

Transient errors are errors that do not exist permanently but occur sporadically. 'Transient' comes from the Latin *transire* and means 'to pass by'. They are caused, for example, by cosmic radiation. Radioactive rays hitting the processor completely at random can, in extreme cases, change the contents of registers or memories, or simply cause the processor to deviate from its specified behavior. These errors can occur despite faultless hardware and software, are rare, unpredictable, and non-reproducible. In addition, it is impossible to foresee what effect they will have. In short, they are the ultimate horror of any software or system developer.

So, is there any way of becoming the master of such errors?

The lock-step concept at least tries. From a software perspective, a lock-step multi-core is a single-core processor. The code is executed by two cores simultaneously and their results are compared by the hardware itself. 'Simultaneous' is, however, not to be understood literally. The execution is delayed on the second core by a few clock cycles so that any sporadic radiation cannot hit the two cores while they are processing the same command.

The Texas Instrument TMS570 [33] serves as an example for a lock-step design. It implements two ARM Cortex R4F cores that are implemented as lock-step cores. The various measures that are used to at least detect transient errors include:

- Time offset during execution (as already described).
- Different units for generating the system clock of both cores.
- The two cores are mirrored and rotated by 90 degrees in the silicon to counteract systematic multiple hardware failures (e.g. due to production errors).

- Large spacing of the cores on the silicon (at least 100 μm).
- Potential guard ring around each core.
- Unit for error handling in the case of deviating results from the cores.
- Self test built into the processor.

7.1.4   Infineon AURIX: Homogenous, Heterogeneous, and Lock-Step

To finalize the topic of 'multi-core processor types', a look at the Infineon AURIX that has been mentioned several time. It combines all types onto one chip, as shown in Figure 73.

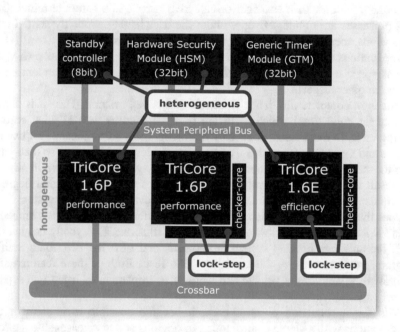

Figure 73   Infineon AURIX multi-core: homogeneous, heterogeneous, and lock-step

7.2   Different Types of Parallel Execution

In the introduction to the chapter "Multi-core", we have already mentioned the three different types of parallel execution, namely:

- Application parallelism.
- Function parallelism.
- Instruction parallelism.

These three types of parallel execution are explained in more detail below.

7.2.1   Application Parallelism

When developing PC software for Windows or Linux, a developer will rarely think about how many processors the PC will have upon which the software will be executed. Users expect the software to run on both a dual-core Intel i3 and a 64 core AMD Threadripper (although obviously with very different performance, if that matters at all). While there are constructs in most programming languages that support multithreading—the newer C++ standards have, in particular, been massively extended here—this does not change the fact that it is mainly the operating system that decides which thread runs on which core at runtime. After all, the one application the developer writes is only one of many that are active at the same time, or at least have been started. You only have to look at the process list in the Windows Task Manager or call `top -n 1` in a Linux terminal to see what processes are running. Even without the user having explicitly started many applications, there will usually be well over 100 processes running.

Applications are developed isolated from one another, each with its own virtual address space and the assumption that they are also largely independent of one another in other respects.

This is in complete difference to classic embedded systems. Here it is clear in advance (at build time) which hardware is to be used as well as what software will run on it. In most cases, not even dynamic memory management (for example `malloc` and `free` under C, and `new` and `delete` under C++) is used, so that—apart from the stack—even the memory addresses of all symbols are known in advance. The result is significantly more predictable systems and a much higher degree of determinism—exactly what you need for safety-relevant and reliable systems.

Does this mean that there is no application parallelism in the classic embedded arena? Not quite, as the following example will illustrate. Everybody has heard of ABS, the anti-lock braking system in vehicles, and ESP, the Electronic Stability Program, made famous by the legendary 'Elk Test'. Both of these features affect the braking system of the vehicle and were often implemented using two separate control units.

With the introduction of multi-core, it became possible to implement both functionalities with a single control unit. An obvious and also reasonable approach was to run the ABS on one core and the ESP on another core of the processor (see Figure 74). The communication that was previously implemented over CAN could now be implemented using shared memory. Both applications run in parallel on different cores of a multi-core processor.

This example demonstrates a sensible use of applications that are both executed in parallel and as independently as possible, and yet such projects are the exception rather than the rule. In the classic embedded arena, application parallelism rarely plays a role.

Somewhere between classical embedded on the one side, and typical PC software on the other, the arena of POSIX-based embedded systems is located. In projects that fall into this category at least the hardware is defined and, with respect to the software, there is less freedom than with PC software.

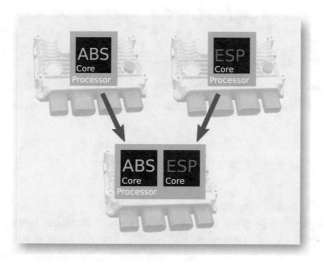

Figure 74 Converting two single core controllers into one multi-core controller

7.2.2   Function Parallelism

As just mentioned, largely independent *applications* executing in parallel are characteristic of PC software, smartphones, tablets etc. In comparison, function parallelism is seldomly seen in High Performance Computing. However, rendering in graphics processing is an example of massive parallelism of a function.

In the classic embedded arena, on the other hand, closely interlinked functions are used. Very often these functions are developed for execution on a single processor core (single-core) and, when switching to multi-core, massive problems are encountered. The close interleaving used does not allow the code to be simply torn apart and distributed over several processor cores. The example that follows should illustrate this.

7.2.2.1 Switching from Single-Core to Multi-Core: Example 'Bubblesort'

The Bubblesort sorting algorithm is simple, commonly known, and can be seen in Listing 27. In this case the function is passed a pointer to an array of `unsigned int` values. The values of the array are sorted in ascending order by the function.

Listing 27 Simple bubblesort implementation

```
1  void BubbleSort(unsigned int* s, unsigned int size)
2  {
3      unsigned int i,j,temp;
4      for(i=1; i<size; i++) {
5          for(j=0; j<(size-i); j++) {
6              if(s[j] > s[j+1])
7              {
8                  temp    = s[j];
```

```
9              s[j]   = s[j+1];
10             s[j+1] = temp;
11           }
12         }
13      }
14 }
```

For an array with n elements the number c of inner loop iterations is calculated as follows.

$$c = \sum_{i=1}^{n} (n - i) = \frac{n^2 - n}{2} \tag{17}$$

This number is decisive for the CET of the algorithm.

For a single-core project, using the BubbleSort function is very easy. It is called, takes some time to complete its task, and then the array is sorted. Figure 75 illustrates this.

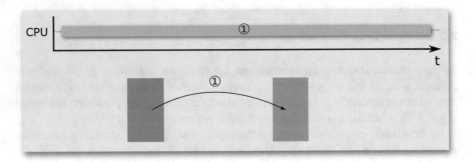

Figure 75   BubbleSort sorting algorithm on a CPU

But how does the use of this function change if it is to be parallelized and distributed across several cores of a multi-core processor? It is easy to see that a function as simple as the BubbleSort sorting algorithm cannot be easily parallelized. The few lines that are sufficient in a single core environment must be significantly extended for multi-core use.

If the array is divided into several subareas, and each CPU is assigned a subarea for sorting, you get several sorted subareas of the array. However, this is not yet the solution. The subareas must then be reassembled to form a sorted overall array. If there are more than two cores, this merging must be undertaken in several steps. Figure 77 illustrates this for three equally sized subsections that are sorted in parallel by three CPUs.

Merging two already sorted arrays can be implemented with the function MergeSortedArrays from Listing 28. In contrast to the BubbleSort function, the number c of loop iterations or copy operations is not proportional to the square of

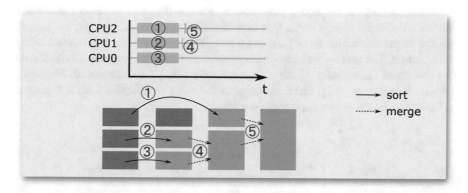

Figure 76 Parallelized sorting algorithm, equal distribution

the number of elements in the array (as is the case with the BubbleSort algorithm).
Instead, it is identical to this number and thus linear: $c = n$.

Listing 28 Merging of already sorted arrays

```
1   void MergeSortedArrays( unsigned int* d,     /* destination */
2                           unsigned int* s,     /* source */
3                           unsigned int size1,
4                           unsigned int size2 )
5   {
6       unsigned int p;
7       unsigned int i1 = 0;
8       unsigned int i2 = size1;
9       for (p=0; p<size1+size2; p++) {
10          if (s[i1] < s[i2]) {
11              d[p] = s[i1++];
12              if (i1 == size1) {
13                  /* reached end of section 1,
14                     so simply copy the rest */
15                  while (i2 < size1+size2) {
16                      d[++p] = s[i2++];
17                  }
18                  break;
19              }
20          } else {
21              d[p] = s[i2++];
22              if (i2 == size1+size2) {
23                  /* reached end of section 2,
24                     so simply copy the rest */
25                  while (i1 < size1) {
26                      d[++p] = s[i1++];
27                  }
28                  break;
29              }
30          }
31       }
32   }
```

When looking at the code, you quickly realize that you shouldn't make the subareas the same size in order that two slightly smaller subareas can be merged while a larger one is still sorting. Above a certain size, the square dependence of the BubbleSort function on the number of elements to be sorted becomes so significant that the linear dependency of the merging hardly plays a role anymore. Figure 77 displays this impressively. The first subset of the array contains half of the elements of the array, the other two each contain a fourth.

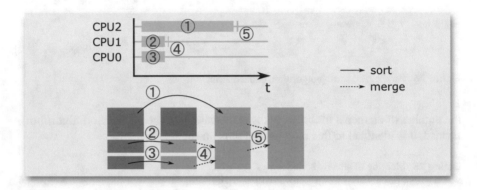

Figure 77 Parallelized sorting algorithm with unequal distribution

For each of the Figures 75, 76, and 77, an array with $n = 1200$ elements was considered. The length of the bars representing the execution of code is equal to the number of loop iterations for each action. When calling BubbleSort in the single-core example (Figure 75), there are $c = \frac{n^2-n}{2} = 719,400$ loop iterations. For the equally distributed approach as shown in Figure 76 it is three times $c = 79,800$, and for the asymmetric approach from Figure 77 it is once $c = 179,700$ and twice $c = 44,850$. With these values the merging of the subareas hardly plays a role anymore and they are only visible as narrow bars in the diagram.

The example for the parallelization of a function was deliberately chosen to be very simple, but the knowledge gained from it can be transferred to more complex algorithms, such as the recognition of objects in a video stream. This initially very simple algorithm for sorting numbers suddenly became much more complex when ported to multi-core. Furthermore, we did not even take into consideration that the CPUs need to be synchronized with each other. After all, we need to ensure that the merging only starts when the affected subareas have been completely sorted. We also failed to address to which memory (shared memory or local CPU memory) the array should best be allocated.

It quickly becomes clear that the parallelization of even a simple function can become very complicated very quickly, and that a simple resolution for this challenge is rarely available. Another aspect has also become clear: the automated transformation of code developed for single-core to a multi-core environment is doomed to failure from the very beginning. It was already highlighted in the introduction that so-called 'C to C compilers' that make this promise have failed to provide proof in everyday project work.

So, what conclusion can we now draw? Should we give up the parallelization of functions in embedded software from the outset? Not at all. If functions are designed to be processed in parallel at the design stage, the use of multi-core processors has great potential. However, if a function is only available at C code level, it is already too late.

This consideration may point to the future path for multi-core. At a high level of abstraction, a functionality is developed or used based upon a model. Then, at the time of code generation, the code generator knows precisely the target system, i.e. it knows which processor is selected and how many cores it has, which operating system is in use, which other software components run on the system, etc. It then generates code that is optimized for use in this environment. The developer does not have to deal with the details. For the example of sorting used in this section, this would mean that in a presumably graphical programming interface, the developer simply feeds his array into a 'sort function block'. Done. Everything else, i.e. how many cores the sorting takes place on, which core is assigned which share, how the cores synchronize with each other, which memories are used, which sorting algorithms are used, etc., is left to the code generator.

Such a development environment is still a dream of the future in 2020. But, if a provider of model-based development tools were to seriously address the field of embedded software in this sense, it would be sure to have a decisive and unique selling point.

7.2.2.2 Duplicate Parallelism vs. Pipeline Parallelism

Typically, a specific functionality is implemented in such a way that several steps are processed one after the other. In the following example, a functionality consists of two such steps: a filtering A in the first and a calculation B in the second step. The execution of both steps together takes longer than one millisecond (see Figure 78).

Figure 78 Splitting a functionality into two parts: A and B

The functionality needs to be executed every millisecond. It is immediately obvious that this is not possible on a single CPU, since the execution of the combination of A and B already takes more than a millisecond. So, let's assume a dual-core processor is available for the code execution. There are basically two approaches for distributing the A blocks and B blocks to the two cores of the dual-core processor. One is *duplicate parallelism* as shown in Figure 79. The A/B combination is executed alternately on one core and then the other.

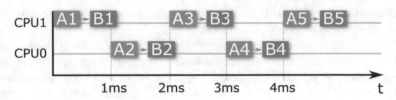

Figure 79 Parallelization through multiplication (duplicate parallelism)

The situation is different with *parallelization through serialization* (pipeline parallelism), which can be seen in Figure 80. Here, part A is always executed on CPU1 and part B on CPU0.

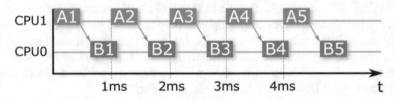

Figure 80 Parallelization through serialization (pipeline parallelism)

Which approach is better, multiplication or sequencing? Duplicate parallelism or pipeline parallelism? This question cannot be answered universally, as it depends on the situation. The following considerations should help in the selection process.

Advantages of Parallelization Through Multiplication

- The data exchange between parts A and B take place locally on the respective CPU. Therefore, no communication across core boundaries is required. Such 'cross-core' communication should always be minimized.
- The scheduling on CPU0 is more deterministic and more predictable. An A/B combination is executed every 2 ms. With parallelization through sequencing, on the other hand, the execution of part B is triggered by the completion of part A. Depending on the runtime of part A, this could occur sooner or later.

Advantages of Parallelization Through Sequencing

- Executing part A only on CPU1 and part B only on CPU0 usually results in a more efficient use of the program cache. In general, if less-different code is executed on one CPU, the cache is better utilized.
- In heterogeneous multi-core processors, it may be possible to distribute code in such a way that some parts of the code run particularly well on a particular core. An example is the Infineon AURIX whose performance cores (recognizable by the 'P' in their core designation) can execute floating-point arithmetic

significantly faster than cores tuned for efficiency (those with an 'E' in their designation). Assuming that part A in the above example uses floating-point arithmetic intensively, while part B does not do so at all or only very little, then part A would be assigned to a 1.6**P** core and part B to a 1.6**E** core.

- It is assumed that the data filtered by part A must be received first. If this reception is always done on the same computational kernel, this can provide another advantage. If all communication with the outside world is handled by this core, the communication stack will run exclusively on this core. With regard to cache and other memory usage, this approach has considerable potential for optimization.

7.2.3   Instruction Parallelism

The parallel execution of commands at hardware level has already been explained in Section 2.6. In connection with the topic 'Parallel Execution' it should be pointed out again that a pipeline does exactly that which its name describes: process several commands in parallel. The CPU does this on its own without any further input required. However, its efficiency-enhancing effect can be impaired by unfavorable jump instructions, as Section 2.6.1 has demonstrated.

The following list offers a few measures for 'pipeline friendly' software:

Avoidance of function calls   A classic example of the cause of unnecessary function calls is the introduction of wrappers, i.c. an adapter layer that adapts the interface of a software component to another interface. This is often used when old code is integrated into a new environment and the new environment requires a similar but different interface. If the functions of the old code are now embedded in otherwise empty functions to meet the new interface requirements, a further function call is added to each function without necessity. It becomes particularly disadvantageous if this undertaken several times, i.e. a wrapper is built around the wrapper of a wrapper.

Instead, other mechanisms should rather be used to map one interface to another. The following options are available, the use of which increases the overall pipeline efficiency and not just in conjunction with the use of wrappers.

- Macros ("`#define` ...")
- `inline` functions

Avoiding interrupts   Similar to the function calls just discussed, avoiding interrupts actually concerns avoiding *unnecessary* interrupts. Avoiding interrupts in general on embedded systems is mostly neither possible, nor useful. However, interrupts are often used to signal the reception of data which will be processed later. Instead of interrupts, it may be possible to simply query (poll) at the beginning of processing whether new data is available. Section 3.1.2 already demonstrates this approach with a code example (Listing 11 on page 39).

It's amazing how many interrupts can be replaced by polling without any penalty. By the way, this also reduces the danger of data inconsistencies and ensures the cache is used more efficiently.

Section 8.1.6 will show how to minimize the number of interrupts required for scheduling.

7.3   Data Consistency, Spinlocks

The topic of data consistency was already covered in Section 2.9 in relation to interrupts. The situation in a multi-core environment is similar, so the example introduced in Section 2.9 will be used again.

Unlike before (see Figure 15), the two interrupts now execute on two different cores of a multi-core processor, as shown in Figure 81.

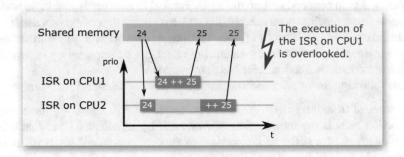

Figure 81   Example of data inconsistency using multi-core

In such situations, interrupt locks are of no help as they only apply to the CPU on which they are triggered. So, how can the unfortunate simultaneous access to shared memory and the resulting data inconsistency be prevented?

One option is the use of *spinlocks*. With their help, one CPU occupies a resource (this is also visible for all other CPUs) and, for the duration of the occupancy, no other CPU can use this resource. In this example, the counting variable counterISR in the shared memory area represents the resource to be protected.

Listing 29 shows the interface for using spinlocks as provided in AUTOSAR.

Listing 29   Spinlocks interface, as defined in AUTOSAR

```
1 StatusType GetSpinlock          ( SpinlockIdType SpinlockId      );
2 StatusType ReleaseSpinlock      ( SpinlockIdType SpinlockId      );
3 StatusType TryToGetSpinlock     ( SpinlockIdType SpinlockId,
4                                   TryToGetSpinlockType* Success );
```

The service GetSpinlock occupies a resource (a spinlock) and the service ReleaseSpinlock releases it again. A simple usage example can be seen in Listing 30.

Listing 30 Use of a spinlock to protect a resource

```
1 GetSpinlock(spinlock);
2 /* This is where the access to the
3    protected resource occurs */
4 ReleaseSpinlock(spinlock);
```

If the resource is already occupied when calling GetSpinlock, the function waits until the resource becomes free. This waiting is implemented by means of a loop within the function GetSpinlock, hence the name (spin equals 'turning in a circle' and 'lock' because, during this time, the execution of other code is excluded). Waiting for a resource is unproductive time and should be avoided, or at least minimized.

The service TryToGetSpinlock will be discussed later. First, a few problematic situations will illustrate the challenges that can occur when using spinlocks.

Our first example is the situation where an interrupt on CPU1 causes a task A on CPU0 to be significantly delayed, even though the interrupt does not access the resource. Figure 82 illustrates this. Shortly after Task B on CPU1 occupies the resource, the Task A on CPU0 also attempts to use the resource. However, no sooner has Task B started using the resource, it is interrupted by an interrupt and must wait before its processing can continue. In the meantime, CPU0 sits 'wasting time' and does not execute any productive code.

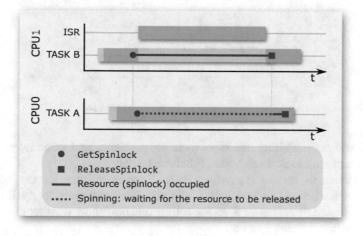

Figure 82 Delay due to an interrupt independent of the protected resource

One possible solution that can be quickly implemented is to enclose resource access with an interrupt lock and release, as shown in Listing 31.

Listing 31 Inhibiting interrupts during resource access

```
1  DisableAllInterrupts();
2  GetSpinlock(spinlock);
3  /* the protected access takes place here */
4  ReleaseSpinlock(spinlock);
5  EnableAllInterrupts();
```

Unfortunately, this results in us jumping from the frying pan and into the fire because now a situation can occur as visualized in Figure 83. Task B is trying to access a resource that was recently acquired by Task A on the other core. As soon as Task B has disabled the interrupts to avoid the problem described above, the interrupt is triggered and must wait to be processed. Although it, again, has nothing to do with the resource, it must still wait for Task A and its resource access. After Task A releases the resource, Task B is now activated and the interrupt must continue waiting until Task B has completed its access to the resource and enabled the interrupts again.

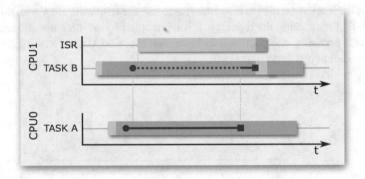

Figure 83 Delay of an interrupt independent of the protected resource

Thus, without inhibiting interrupts there are problems, and when interrupts are inhibited it doesn't look much better. What can be done?

This is where the service `TryToGetSpinlock` comes into play. It is non-blocking, meaning that, whatever the state of the spinlock, it returns immediately. It uses the parameter passed by reference to let the calling function know whether the resource was successfully allocated or not.

The code example below (Listing 32) shows a more skillful use of spinlocks than in the previous two examples. This code should be used instead of a simple combination of `GetSpinlock` and `ReleaseSpinlock` to overcome the problems described.

After disabling the interrupts in line 2, an attempt is made to occupy the resource using `TryToGetSpinlock` in line 3. If this succeeds, the resource can be used in a manner where it is protected and ensured that its use is not delayed by an interrupt—the situation as shown in Figure 82 is thus excluded. If the resource cannot be used because it has already been blocked by other code, the CPU now waits in a loop for the resource to be released. In this loop the interrupts are briefly enabled to allow interrupts to occur while waiting for the spinlock. This prevents a situation as shown in Figure 83.

Listing 32 Best practice use of `TryToGetSpinlock`

```
1  TryToGetSpinlockType success;
2  DisableOSInterrupts( );
3  (void)TryToGetSpinlock( spinlock, &success );
4  while( TRYTOGETSPINLOCK_NOSUCCESS == success )
5  {
6      EnableOSInterrupts( );
7      /* Interrupts that occur will be handled here */
8      DisableOSInterrupts( );
9      (void)TryToGetSpinlock( spinlock, &success );
10 }
11 /* Access to the protected resource
12    occurs here */
13 ReleaseSpinlock( );
14 EnableOSInterrupts( );
```

Can this code be optimized even further?

The loop calls the service `TryToGetSpinlock` very often during the waiting period. Since the status of the spinlock must be made available to all cores of the processor, the corresponding internal operating system variable will be in shared memory. The high frequency calls of `TryToGetSpinlock` will therefore lead to many accesses on the shared memory bus. The probability of access conflicts on the bus, with their corresponding delays, increases enormously. This will most likely slow down the core that is currently accessing the resource—obviously delaying its release.

To counteract this, one could insert no-operation commands (such as a 'nop') into the loop body that have no other purpose than to reduce the frequency of the `TryToGetSpinlock` calls. This measure is also not entirely without disadvantage as, in most cases, it results in an additional delay when waiting for the resource to be released.

Admittedly, this last approach, that of inserting 'nop' instructions, is rather extreme and is thus not used that often. The reason it is mentioned at this point is because it highlights, quite beautifully, several different aspects of embedded software and promotes an understanding of how microprocessors work in the context of embedded software timing.

7.3.1 The Ideal Solution for Ensuring Data Consistency

As already mentioned in connection with the interrupts in Section 2.9, the best way of ensuring data consistency is the one you do not need. Applied specifically to this example, it can be seen as an alternative implementation. Listing 33 shows a simple approach that works for both the case of data inconsistency in interrupts (see Figure 15) and the situation where two CPUs of a multi-core processor are competing (see Figure 81). Each interrupt simply uses its own counter and, whenever the sum of the executions is required, the sum is calculated at the moment of the query. Even if this query is interrupted by one of the interrupts, you still obtain a correct value.

For the sake of clarity, it was decided not to intercept any overflow of the counters and the total in this example. In a real project, this would have to be implemented if there was a danger that could occur from an overflow.

Listing 33 The ideal solution where extra protection is not necessary

```
1  unsigned int counterISR_low_prio = 0;
2  unsigned int counterISR_high_prio = 0;
3
4  void ISR_low_prio (void) __attribute__ ((signal,used));
5  void ISR_low_prio (void)
6  {
7      _enable(); // globally enable interrupts
8      counterISR_low_prio++;
9      DoSomething();
10 }
11
12 void ISR_high_prio (void) __attribute__ ((signal,used));
13 void ISR_high_prio (void)
14 {
15     _enable(); // globally enable interrupts
16     counterISR_high_prio++;
17     DoSomethingElse();
18 }
19
20 unsigned int GetCounterSum(void)
21 {
22     return counterISR_low_prio + counterISR_high_prio;
23 }
```

7.3.2   The Costs of Ensuring Data Consistency

The costs of data consistency assurance will be illustrated by means of a retrospective view. Towards the end of the last century, the majority of embedded software was written by hand and a global interrupt lock was found in those places where data consistency had to be ensured. This was implemented either directly using assembly language code (such as __asm(di) and __asm(ei)) or using special commands (such as __disable() and __enable()). Such C language extensions, for which the compiler translated each into exactly one machine instruction, are compiler- and processor-specific functions. The cost of data consistency assurance, requiring just one machine instruction each, was extremely low. When we talk about costs here, this is to be understood in terms of runtime. In many projects, the interrupts were disabled and enabled several thousand times per second. With the low overhead involved, this was acceptable.

With the introduction of operating system interfaces for managing interrupts, such as those offered by OSEK, these costs more than doubled. Instead of a single machine command for disabling and enabling, a function call was added to each. This was already painful but, in most cases, still bearable.

A few years later, safety operating systems were introduced on a larger scale and, when disabling interrupts, it was now necessary to additionally check whether the context just executed was even authorized to disable interrupts. The runtime overhead thus increased again by a huge amount.

After all, more and more projects were using multi-core processors and here—as discussed in detail above—simply blocking interrupts was no longer sufficient to ensure data consistency. Instead, spinlocks are used with the result that, when the resource to be protected is accessed simultaneously, two or even more CPUs are now affected and sometimes do nothing else but wait.

Figure 84 illustrates the different protection mechanisms and their costs.

If code from an old single-core project is ported to a multi-core platform—one which implements the 'thousands of simple interrupt locks per second' approach—it is often the case that a simple 'search and replace' is performed to replace the simple and very inexpensive interrupt locks with very expensive spin locks. The result is typically a hopelessly overloaded multi-core processor and correspondingly long-faced developers.

7.4   Cloning of Memory Addresses

Infineon introduced a feature during the development of the TriCore AURIX architecture that offers optimization potential in the implementation of software for multi-core while also significantly simplifying the porting of single core software to multi-core. This is described as 'cloning memory addresses', the details of which can be found in the AURIX manual under the chapter "Local and Global Addressing".

Basically, it is a mechanism for efficient access to CPU local memory. Each AURIX CPU has a local program memory called PSPR and a local data memory called DSPR. PSPR stands for "**P**rogram **S**cratch **P**ad **R**AM" and DSPR for "**D**ata **S**cratch **P**ad **R**AM" (see Figure 4). Here we will only review the DSPR as the PSPR functions in a similar manner.

Figure 85 shows the addresses where the DSPR memory of the respective CPUs can be accessed. Infineon likes to use the underscore notation "_" between the upper 16 bits and the lower 16 bits of the address for better readability.

For clarity, only three CPUs are shown in the Figure 85. For AURIX processors with six cores, the DSPR of CPU3 is located at the global address 0x4000_0000, that of CPU4 at address 0x3000_0000, and that of CPU6 at address 0x1000_0000. The numbering of the six CPUs is somewhat confusing as there is CPU0 to CPU4 and then CPU6. Section 8.3.3 will show how to efficiently handle the core IDs on six-core AURIX processors. Regardless of the numbering, each CPU has its own DSPR that occupies a certain range in the global linear address space.

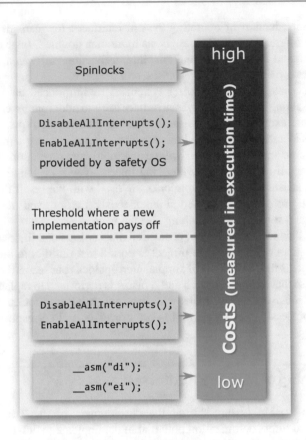

Figure 84   Cost (measured in runtime) for ensuring data consistency

So far, this largely corresponds to the structure of other multi-core processors. The special feature is that each DSPR is also *additionally* accessible via an address range starting at 0xD000_0000. When this area is accessed, the DSPR is always used of the CPU upon which the access was executed. For example, if CPU1 accesses the address 0xD000_00C4, the DSPR of CPU1 is used. In the global linear address space this access corresponds to an access to the address 0x6000_00C4.

What benefits can be derived from this structure? This quickly becomes clear when you look at code that is to be used on several of the cores. A real-time operating system is a good example of such code. Every operating system has various internal variables, such as the one that identifies the task currently running. For a single-core operating system, the definition and use of this variable might look as in Listing 34.

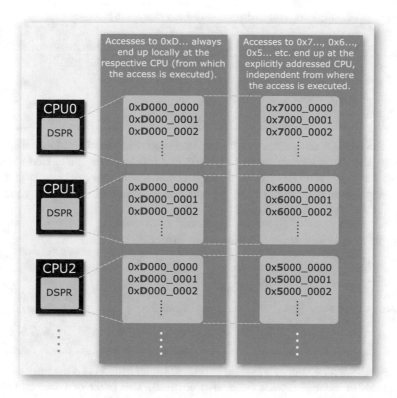

Figure 85 Access logic when using memory address cloning

Listing 34 Code snippet of an operating system for single-core processors

```
1  unsigned int runningTask;
2
3  void someOSfunction(void)
4  {
5      ...
6      runningTask = ... ;
7      ...
8  }
```

Let us assume that the operating system is now to be used on a multi-core processor in such a way that it runs in isolation on each CPU. A separate instance of the variable runningTask must therefore be used on each CPU. The program code itself can, or should be, used by all CPUs. Listing 35 shows an implementation often found in generic multi-core code. In this context, 'generic' means that the code can be translated to different multi-core processors—even those that do not support memory address cloning. The variable runningTask has now become an array and each CPU of the processor gets its own element of this array. Upon each access, the currently executing core must ensure that it uses 'its' element.

Listing 35 Code snippet of an operating system for any multi-core processor

```
1 unsigned int runningTask[NOF_CORES];
2
3 void someOSfunction(void)
4 {
5     ...
6     runningTask[GetCoreId()] = ... ;
7     ...
8 }
```

On the AURIX this access can be implemented using the CPU-local DSPR by cloning. The compiler only needs to be told that this variable is a cloned variable. Listing 36 shows how this can be done. It is noticeable that the code is identical to the single-core code from Listing 34 with the exception of the memory qualifier __clone.

Listing 36 Code snippet using memory cloning on AURIX

```
1 __clone unsigned int runningTask;
2
3 void someOSfunction(void)
4 {
5     ...
6     runningTask = ... ;
7     ...
8 }
```

The TASKING compiler was used in this example. Instead of a memory qualifier a `#pragma` or compiler switch can also be used. The latter of these allows the unchanged single-core code to be used on a multi-core AURIX, which can simplify the porting.

As well as easier code handling, access to cloned memory is also more efficient. There is also no need to query the core ID, which additionally saves runtime and code memory.

7.5   Summary

In Chapter 7 we started with an introduction to the different types of multi-core solutions that are available. There are a range of different types including heterogeneous, homogeneous, and lock-step. It is not always possible to classify a microprocessor into one or the other category, as often several are combined on one piece of silicon.

After the hardware architecture the different types of parallelism were examined:

- Application parallelism.
- Function parallelism.
- Instruction parallelism.

While parallelization of applications and parallel execution at the instruction level is quite easy to implement, parallelization of functions often presents developers with major challenges. When it comes to making existing code at this level multi-core capable, reimplementation is often the best way to go.

Perhaps the parallelization of functions will gain in importance in the future. This will be the case when the code generators of functional modeling tools are able to generate efficient code for a given system by exploiting parallel execution. Until this is achieved, it is the responsibility of the system architects together with the function developers, and integrators with the appropriate knowledge, to build efficient systems.

The topic of data consistency was covered again, this time in relation to multi-core. Spinlocks were described in detail, including instructions for their efficient use.

The chapter was rounded off with a look at the concept of memory address cloning. This allows both the efficient implementation of code executed by multiple CPUs and can simplify the porting of single-core code to multi-core devices.

Timing Optimization

8

Timing optimization follows a strictly 'top-down' approach. This means that the scheduling level is analyzed and optimized first, followed only then by optimizing at the code level. If one were to start directly at the code level, they would run the risk of spending a lot of time on code optimizations that improve the runtime at a point that is not critical at all.

The optimization of memory usage is, in a way, orthogonal to the mentioned levels. A memory optimized for timing distributes the symbols over the available memory in a way that minimizes the total CPU utilization. Of course, it must still be guaranteed that memory overflows at compile time and runtime are excluded, and that possible safety requirements are taken into account.

8.1 Timing Optimization at the Scheduling Level

Unfortunately, there are only a small number of measures for runtime optimization at the scheduling level that can be worked through using a checklist approach. The greatest potential for optimization at the scheduling level lies in project-specific parameters. This means the basic distribution of the application across the different cores of a multi-core processor, the configuration of the operating systems, and the distribution of functions or runnables to the tasks, etc. In other words, the entire project-specific *timing design* certainly has the greatest influence on the efficiency of the scheduling. Timing design naturally takes place at an early stage of the project.

In this phase, scheduling simulation and static scheduling analysis can be very beneficial. The techniques are explained in detail in the Chapter 5. They can be used to weigh up different approaches to timing design against each other, or to perform an optimization. During optimization, both the optimization goals and the degrees of freedom can be defined. Section 5.10 has examined this aspect in more detail.

The following sections are intended to support a timing design achieving efficient scheduling and, as far as possible, each should be considered. 'Should' here means

© The Author(s), under exclusive license to Springer Nature Switzerland AG 2021
P. Gliwa, *Embedded Software Timing*,
https://doi.org/10.1007/978-3-030-64144-3_8

that there will be good reasons for many projects to consciously *not* implement one or the other aspect.

Some of the aspects can also be used in later project phases to optimize the runtime behavior.

8.1.1 Prevention of Cross-Core Communication

The distribution of functionality across the different cores of a multi-core processor should be done in a way that minimizes communication across the core boundaries.

8.1.2 Separation of Computationally Intensive Code and Interrupts

As many interrupts as possible, or even all of them, should be handled by one core, while computationally intensive code sections should be deployed to another core. This distribution promotes efficient use of the cache and pipeline on the core that runs the compute-intensive sections. Splitting functionality across the different cores of a multi-core processor in this manner helps to optimize the overall use of both the pipeline and cache.

8.1.3 Avoiding the Use of ECC Tasks

This aspect is only relevant for OSEK/AUTOSAR CP projects. Section 6.2 pointed out that most RTE code generators use an ECC setup by default. The section also described the considerable disadvantages of this approach. Whenever cyclic runnables need to be scheduled, a configuration with a single cyclic BCC1 task per period is a much better choice than a non-terminating ECC task handling all periods.

8.1.4 Sensible Use of Heterogeneous Multi-Core Processors

In heterogeneous multi-core processors the division should be made in such a way that the computationally intensive parts of the software are assigned to the most powerful cores. The Infineon AURIX, for example, offers 1.6**P** and 1.6**E** cores. The 'P' stands for performance and the 'E' for efficiency. The same code can have significant runtime differences depending on whether it is running on a 1.6**P** or 1.6**E** CPU. Experience has shown that speculation hinders such decision making. The most efficient way to decide which code should be executed on which core is to undertake runtime measurements to substantiate the decision. For safety-relevant projects, a static code analysis can also reliably determine theoretical worst case runtimes — but not average runtimes.

8.1.5   Avoiding the Necessity for Mechanisms That Ensure Data Consistency

The topic 'data consistency' was covered in the Section 2.9 in general and in the Section 7.3 for multi-core. Section 7.3.2 addressed the fact that the cost of ensuring data consistency has increased as systems have become more complex, and that the best concepts are those that eliminate the need for explicit data-consistency mechanisms.

The tricky thing about data-consistency mechanisms is that the user is not usually aware that they are used in large numbers. At the modeling level, various communication mechanisms are implemented. These are clean and correspond to good development practice. Only later, when the code generators translate the model into C code, do they analyze the system and, whenever there is a risk of data inconsistency, implement appropriate data-consistency mechanisms. These almost always consume additional RAM, flash, and runtime.

Ideally, the system — especially the operating system — will be configured in such a way that data-consistency mechanisms are not even required, and thus no additional code will be generated. The following optimization measures support this approach.

- Avoid preemptive interruption as a matter of principle by using the same priorities or priority groups whenever possible.
- Avoid preemptive interruption as a matter of principle by using cooperative multitasking (scc Scction 3.3).
- If preemptive interruption cannot be avoided, it is sometimes useful to divide the preemptive task or (preemptive) interrupt into a section that contains the mandatory preemptive parts, and a section that is not preemptively implemented and contains all the other parts. This second part can be implemented as a non-preemptive task, or as code that runs in the background task and is triggered by a flag or another similar mechanism.

The remaining preemptive parts should operate on as little shared data as possible, limiting the amount of data requiring data-consistency mechanisms.

8.1.6   Load Balancing with Optimized Offsets for Periodical Tasks

When several periodic tasks are configured, the question arises as to their temporal relationship to each other. This is set via the *offsets*, a time difference to the 'zero line' of the start of the scheduling or an imaginary zero line. Figure 86 shows a runtime situation with three tasks, all of whose offsets have been configured with 0. At the time when the `Task_1ms` is activated for the second time, `Task_5ms` is still running, but is interrupted. Figure 87 shows the same runtime situation (same periods, priorities, and CETs), with the only difference being that the offsets of the two tasks `Task_2ms` and `Task_5ms` were not set to zero. As a result, the `Task_5ms` is no longer interrupted and the computing load is distributed more evenly over the

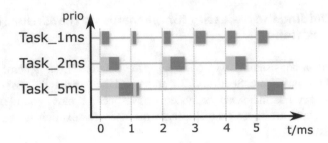

Fig. 86 All tasks have an offset of zero: high load at t=0 ms

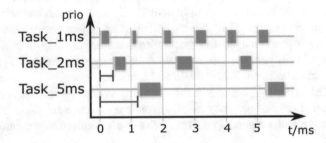

Fig. 87 Only `Task_1ms` has an offset of zero. Result: optimal load distribution

time axis. Furthermore, all IPTs (**I**nitial **P**ending **T**imes, initial wait times) of the `Task_2ms` and `Task_5ms` tasks are lower than before.

So, is it advisable to add offsets to all tasks, except the one with the smallest period, in order to attain optimal load distribution? To answer this question, it is necessary to briefly explain how most operating systems activate periodic tasks. These are implemented using a 'scheduling interrupt'. This scheduling interrupt is triggered at the 'due time' of each task activation. In most cases, the compare interrupt of a timer is used that is configured to trigger at the next task activation time — similar to an alarm clock. If several task activations are due at that time, they are handled in the same ISR (Interrupt Service Routine). In other words, offsets that lead to simultaneous task activations help to reduce the number of scheduling interrupts.

The goal of distributing the computing load as evenly as possible along the time axis contradicts, to a certain extent, the optimization approach of reducing the number of scheduling interrupts by activating tasks simultaneously. What is the optimal solution? For most systems, the following simple procedure can be used.

Hint *The 'fastest' periodic task receives an offset of zero and all other periodic tasks receive an offset that is an integer multiple of the period of the fastest task. This results in the minimum number of required scheduling interrupts. At the same time, this approach leaves enough room for sufficient load balancing.*

Figure 88 shows such a configuration. Six scheduling interrupts are required for all the task activations in the visible time slice. This corresponds to the number of `Task_1ms` instances. The configuration shown earlier in Figure 87 required eleven scheduling interrupts, almost twice as many. Nevertheless, the load distribution shown in Figure 88 remains acceptable.

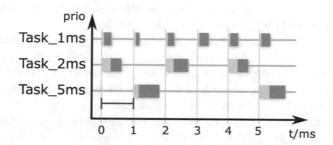

Fig. 88   Minimum number of scheduling interrupts with good load balancing

For the small example shown here with only three tasks, the distribution of the computing load between the configuration without any offsets and the optimized configuration with an offset for the `Task_5ms` is not very large. But this changes quickly when more tasks are added. By the way, Figure 24 already showed the trace of a series production project using optimized offsets. Every activation of a task with a longer cycle time is done together with the activation of the 'fast' 1 ms task.

8.1.7   Splitting of Tasks

Another very simple way to distribute load is to split tasks. Let us assume that an application serves a substantial part of its functionality via the background task. At the same time, larger portions are located in a cyclical task with a period of 5 ms, which results in a correspondingly long runtime. As a result, the code from the background task is not executed during the rather long execution time of the cyclical task, which may be problematic.

A simple solution to the problem is to split the cyclical task into two cyclical tasks with a period of 5 ms and different offsets. This allows a gap between the two tasks to be 'designed in' so that the background task is used more often. With respect to Figure 88, this would mean splitting `Task_5ms` into `TaskA_5ms` with an offset of one millisecond, and `TaskB_5ms` with an offset of 3 ms.

Whether such a division of tasks makes sense for a project depends strongly on the project itself. The additional task also consumes additional resources. The above example has shown a convenient use case, and there are countless others. The decisive factor is to be aware that splitting tasks at the scheduling level during timing design or runtime optimization can be undertaken, and then evaluating this approach for the current project.

8.1.8   Moving Functionality to Less Frequently Executed Tasks

On the one hand, this optimization approach is trivial but, on the other, it can have great pertinence in practice. It is all about questioning whether all cyclically executed code sections could be executed less frequently.

For example, if it is determined that a runnable can be executed once every 10 ms, instead of once every millisecond, without affecting functionality, the runtime requirements of this runnable can be reduced by 90%.

Often, control algorithms operate using a period of 10 ms because, in many cases, this is the period for most of the application's communication. Perhaps the corresponding task can be divided into a section that continues to handle communication every 10 ms, and another section that calculates the control algorithm in a newly created 15 ms task, for example. Before such a measure is taken it is of course necessary to examine, in detail, the time dependencies between the communication code and the control algorithm.

At the beginning of this section it was mentioned that the approach described here refers to cyclical code sections. However, it can also be applied to non-periodic, sporadic code. For example, if an interrupt handles a range of tasks in its ISR, one can question which elements must be executed in the context of the interrupt, and which elements can be removed and processed in the background task.

It is always worth questioning whether code can be executed less frequently without affecting functionality.

8.2   Timing Optimization of Memory Usage

The optimization of memory usage cannot be simply assigned to the scheduling or code level as, in some ways, it affects both levels. The central question in this context is how often symbols are used and relates to both the scheduling (for example, the configured period of cyclic tasks) and the code (for example, the number of loop iterations).

8.2.1   Optimal Use of Fast Memory

In Chapter 2 we saw that microprocessors have both different memories and addressing modes. Some allow fast access, others are slower. The question that arises is: Where should the symbols be located to achieve the highest possible performance of the software? The fast memories are usually quite limited, so there is not enough space for all the symbols. The same applies to the fast addressing modes: the address range for accesses is limited and is usually not sufficient to be able to address all symbols as near.

Again, as a reminder, when we talk about symbols, we are referring to both code symbols (mostly functions) and data symbols (mostly variables). If not defined otherwise, most compilers assign code to the section `.text`, while the data ends up in the sections `.bss`, `.data`, or `.noinit`. Constant data plays a special role because, in one sense, it is data and, in another, it is located in flash memory. By default, such symbols are assigned to the section `.rodata`. Section 1.3.8 has taken a closer look at this topic and explained in detail how symbols are assigned to the different storage and addressing modes using sections.

So, what strategy should be used when you need to specify locations for thousands of symbols and minimize CPU load? Figure 89 answers this question. Small symbols that are accessed frequently are especially suitable candidates for placement in fast memories. If large symbols were assigned to the, typically, very limited fast memories, fewer symbols would fit into these memories in total — with the result that the number of *efficient* accesses would be lower.

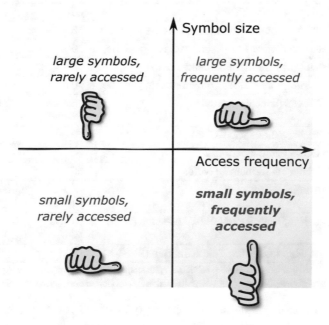

Fig. 89 Which symbols belong in fast memory?

In the following, access times for these symbols will be examined and quantified in a little more detail. Figure 90 shows a table from the manual of a first generation Infineon AURIX. It shows how many clock cycles the CPU has to wait until the access to various memories is complete (stall cycles).

One important aspect is covered in the text of the manual. All of the figures provided refer to situations where no conflicts occur during access. Conflicts during accesses occur, for example, when two cores attempt to access the same memory area at the same time. The memory logic of the AURIX can be configured

On-Chip System Buses and Bus Bridges

Table 2-16   CPU access latency in CPU clock cycles for TC27x

CPU Access Mode	CPU clock cycles
Data read access to own DSPR	0
Data write access to own DSPR	0
Data read access to own or other PSPR	5
Data write access to own or other PSPR	0
Data read access to other DSPR	5
Data write access to other DSPR	0
Instruction fetch from own PSPR	0
Instruction fetch from other PSPR (critical word)	5
Instruction fetch from other PSPR (any remaining words)	0
Instruction fetch from other DSPR (critical word)	5
Instruction fetch from other DSPR (any remaining words)	0
Initial Pflash Access (critical word)	5 + configured PFlash Wait States[1]
Initial Pflash Access (remaining words)	0
PMU PFlash Buffer Hit (critical word)	4
PMU PFlash Buffer Hit (remaining words)	0
Initial Dflash Access	5 + configured DFlash Wait States[2]
TC1.6E/P Data read from System Peripheral Bus (SPB)	4 ($f_{CPU}=f_{SPB}$) 7 ($f_{CPU}=2*f_{SPB}$)
TC1.6E/P Data write to System Peripheral Bus (SPB)	0

1) FCON.WSPFLASH + FCON.WSECPF (see PMU chapter for the detailed description of these parameters).

2) FCON.WSDFLASH + FCON.WSECDF (see PMU chapter for the detailed description of these parameters).

Fig. 90 Excerpt from the AURIX manual with memory access stall-cycles

extensively and, by assigning priorities, you can determine how to handle such conflicts. If a high-priority core reads an entire memory area — such as an array — the delay this causes to another core can be extreme.

By the way, this is one of the reasons why a strict static WCET code analysis is rarely useful in a multi-core environment. For every single access the CPU under consideration makes, the analysis must assume that all theoretically possible conflicts actually occur. This is an assumption that is so pessimistic that the result of the analysis has no practical use. Static code analysis was discussed in detail in Section 5.3.

However, let's get back to the stall-cycles. During the stall time, the CPU cannot perform any other meaningful activity. Thus, the goal of optimized memory usage is to minimize the number of stall cycles for the entire system.

In order to be able to carry out the optimization, the size of the symbols and their frequency of access — or, in the case of functions, the frequency of calls — must be known.

The size of all symbols is very easy to determine and you only need to look at the linker map, i.e. the file created by the linker that provides information about the address of each symbol and its size.

It is more difficult to determine the frequency of accesses and calls. The simplest way to obtain these figures is by measurement or by tracing. For tracing, hardware-based tracing is a good idea (see Section 5.6). In the case of runtime measurement, an approach that makes it possible to instrument the functions or data accesses at runtime can be used (see Figure 57). Thus there is no need to modify the source code and recompile it between measurements. In addition, symbols for which no source code is available at all, perhaps because they were delivered as object code, can also be measured.

In exceptional cases, instrumentation of functions by the compiler itself may be useful. However, this approach brings several problems with it. Functions whose source code is not available cannot be considered. Both the code and measurement overhead is considerable and in many cases — if not most — results in software that can no longer be executed. Ultimately, the software is modified, which means that the functions to be measured no longer correspond to those that were originally intended to be analyzed. Many compiler optimizations, such as leaf-call optimizations (see Section 8.3.6), no longer work and, especially with small functions, the effect of static instrumentation is significant.

8.2.2   Alignment of Data

Alignment describes the offset of a symbol in memory to an imaginary grid that is projected onto memory. An example illustrates this.

A 32-bit processor provides commands to load 32-bit words from memory. However, the processor requires that the address of the data is an integer multiple of four. Why four? Quite simply, memory addresses can be understood (with some exceptions) as positional information for bytes, and one byte contains eight bits, therefore: $4 \cdot 8 = 32$. The situation resembles a four-byte (or 32-bit) grid that is laid over the memory and into which all validly positioned data must fit. On the left side of Figure 91 this is the case, as the data is at address 0x14D0 and fits into the light-blue 32-bit grid. All addresses with 32-bit alignment end with 0, 4, 8, or C.

The processor also provides commands that allow access to data with a size of 16 bits. Whenever a 32-bit word is accessed that is not in the 32-bit grid but in the 16-bit grid in memory, the processor must make two separate 16-bit accesses and assemble the result into the desired 32-bit word. This works, but the resultant access is much slower. The middle illustration in Figure 91 shows this case: the data is at

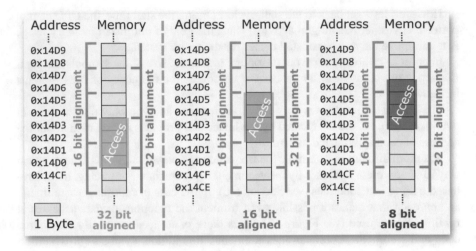

Fig. 91 32-bit, 16-bit, and 8-bit alignment of a 32-bit data symbol

address 0x14D2 and fits into the brownish 16-bit grid. All addresses with 16-bit alignment end in 0, 2, 4, 6, 8, A, C, or E.

If the data is located at an address that does not fit into the 32-bit grid or the 16-bit grid, no 32-bit access is possible. If such data must still be read, the programmer has to implement individual accesses — for example, two 8-bit and one 16-bit accesses — and manually assemble the result. The overhead compared to an access with 32-bit alignment is enormous. The right-hand side in Figure 91 shows a 32-bit word that is misaligned.

Many 32-bit architectures actually function like the example processor described here. 16-bit processors behave similarly: they typically work more efficiently with data in 16-bit alignment. The 8-bit processors have it easier. Only special 16-bit instructions may be subject to alignment limitations with these processors.

How can we influence this alignment of data in memory? Typically, this is done by means of attributes that are assigned to the corresponding output section in the linker script. If a project having runtime problems is not lacking in free memory space, the introduction of alignment for different output sections is a measure that can be implemented quickly. It results in unused gaps in the memory, but some of the memory accesses can be implemented faster. The effect should be checked (like any optimization measure) by tracing or runtime measurement.

It is also the case that the alignment plays an important (and perhaps even more critical) role with data structures (**struct**). The way in which data is packed into a data structure not only influences the memory requirements but also the access speed. If possible, when creating a structure, the individual elements should be grouped so that 16-bit data has a 16-bit alignment relative to the beginning of the structure, 32-bit data has a 32-bit alignment, and so on.

8.2.3   Alignment of Code and Cache Optimization

Machine instructions must be stored in memory with an alignment that corresponds to the instruction length. Most 32-bit processors have a set of efficient 16-bit instructions in addition to their 32-bit instructions. According to the considerations in the previous section, the 32-bit instructions must be stored in memory with a 32-bit alignment, while the 16-bit instructions must be stored in memory with a 16-bit alignment.

Can it be useful to influence the alignment of functions beyond that? This was covered with respect to caches in Section 2.5 in detail. The cache, as a very fast intermediate memory, is organized in cache lines that, in turn, lie across that memory in a manner similar to an imaginary grid. This grid is, however, much coarser than the grids discussed in the previous section. Both Infineon's AURIX and the e200 core of Freescale's power architecture have an L1-cache with cache lines of 32 bytes (256 bits).

A comprehensive cache optimization can be very costly. It is difficult to determine where each function has to be located in memory in order to make optimum use of the cache and this may change with each software release.

If memory space is not a problem, a quick fix can be made and an alignment can be specified for code sections that correspond to the size of a cache line. A software release built in this way will require much more memory than before, since there will now be many unused gaps in the program memory. This can pay off, especially with small functions that fit completely into one cache line and, as a result, no longer occupy two cache lines, resulting in a more efficient use of the cache.

Once again, the effect of this measure is very difficult to estimate in advance, so checks must be made afterwards by means of tracing or runtime measurement.

For both code and data alignment, runtime optimization usually comes at the expense of memory requirements. So, you have to decide between the two optimization goals of 'shortest runtime' (optimization for speed) and 'lowest memory requirements' (optimization for size).

8.3   Timing Optimization at Code Level

As mentioned before, runtime optimization at code level should only be approached after the scheduling level has been considered. Otherwise, it is possible that a lot of effort will be spent optimizing code that is only executed at times when timing is not critical. Conversely, this does not mean that it is absolutely necessary to determine all the time at which code is called for *all* functions. Generally speaking, the interesting candidates (functions) for code optimization can be divided into two categories:

1. Functions that are called from few or even only one place, but have a high runtime requirement.

2. Functions that are called — often from many different places — very often.

An example of a function in the first category would be a complex calculation that takes place within a single large function. The second category usually includes comparatively small functions such as mathematical functions (`sin`, `cos`, `exp`, `sqrt`, etc.) or functions for adapting the data type (type casting). Such type conversions sometimes occur implicitly without the user or the code generator having made a function call. Listing 37 shows a function that contains an implicit type cast. The resulting assembly code generated with the TASKING compiler for the Infineon AURIX is shown in Listing 38. It shows that the type conversion is implemented by calling the function `__d_itod`, which is provided as part of a function library by the compiler vendor.

Listing 37 Example of an implicit type conversion

```
1 int i = 42;
2 volatile double f;
3
4 void main(void)
5 {
6     f = i;
7 }
```

Listing 38 Function call resulting from implicit type conversion

```
1 main:
2   ld.w   d4,i
3   call   __d_itod
4   st.d   f,e2
5   ret
```

The functions of the second category are especially at risk of not being considered at all if runtime analysis is not performed systematically. 'Systematic' here means that, in the best case, the call frequency is determined for all functions. This is not the same as the number of calls in the software, which can easily be determined statically by code analysis. Additionally, the call frequency cannot easily be derived from the number of calls in the software. To do so, one would have to know both the call frequency of the calling function as well as the number of loop iterations if the function call is within a loop.

As mentioned earlier in the section on "Runtime Optimized Memory Usage" (Section 8.2), the easiest way to obtain the call frequency is to use runtime measurement or tracing.

In addition to the call frequency f_F of a function F, its net runtime CET_F is the other decisive parameter for optimization at the code level. If the distribution is sufficiently even, the average values are used for both parameters and thus the average CPU load U_F used by the function can be calculated.

$$U_F = f_F \cdot CET_F \qquad (8.1)$$

Example: from a flow trace an average call frequency $f_F = 23200\,\text{Hz}$ and an average net execution time $CET_F = 570\,\text{ns}$ is determined for the sine function sin in some radar software. This results in the function occupying 1.32% of the available time of the CPU.

Especially in a situation where a CPU is totally overloaded, it has proven to be useful to systematically measure all functions and to create a list that includes the responsible developer together with the CPU load U_F resulting from their functions (see Figure 92).

It is also useful to add another aspect to the list that reflects the expected success of the optimization measure. Some of the functions will only exist as object code and will not be easy to replace, making optimization difficult and costly. Another section may have been optimized in detail in the past, so that further optimization does not promise any significant improvement. For yet another section, it could be immediately obvious that it offers a lot of potential for optimization because, for example, a first look at the code has revealed that inlining can be applied. This group of functions can then be regarded as 'low hanging fruit', i.e. with only little effort, significant improvements can be achieved.

All these aspects can be combined into an 'Expected Success' factor that, multiplied by the CPU load, provides a measure of how worthwhile it is to attack the optimization of a specific function. The table shown in Figure 92 labels this measure 'Priority' and it is calculated as the normalized product of 'Expected Success' and 'CPU load', generating values between zero and one hundred percent as shown.

Whether one follows this suggestion or develops one's own approach to code optimization, the decisive factor is that simple and robust criteria are defined with as little effort as possible to control the procedure for code optimization. Since embedded systems quickly result in tens of thousands of functions, those who base their code optimization strategy on guesswork are not working efficiently.

8.3.1   Optimization of Small, Frequently-Called Functions

In this section, we will look at functions that fall into the second category, i.e. small functions that generate a considerable computing load due to being called often.

Once a function has been identified whose optimization appears worthwhile, the question arises as to how the optimization can be achieved. If the source code is available, it should be analyzed. The best way to do this is to look at the source code and the assembly code generated by the compiler at the same time. The experienced code optimizer can often be recognized by the fact that they can display the source code on one half of the screen and the corresponding assembly code on the other. It is not always easy to assign source code lines to assembly code, especially when the compiler reorders the machine instructions to improve the runtime by cleverly using the pipeline. As a result, the sequential instructions of the source code can appear spread wildly in the machine code.

Nevertheless, a look at the assembler code is worthwhile in most cases. For example, you can see immediately whether function calls are taking place. The

Code optimisation plan						
Project:			Responsible:		File name:	
Software version:			Phone/e-mail:		Version:	
Function	Ø Call frequency f [Hz]	Ø CET [ns]	Ø CPU load U	Expected success	Priority	Comment
mtl_u32_Add_u32_u32	67575	756	5,109%	40%	100%	Improve implement. of saturation
Com_PackSignal	50223	488	2,451%	80%	96%	Comes with two calls: inline?
mtl_s32_Add_u32_s32	55288	521	2,881%	40%	56%	Improve implement. of saturation
mtl_s32_Add_s32_s32	41432	677	2,805%	40%	55%	Improve implement. of saturation
sin	23200	570	1,322%	80%	52%	Re-implement (don't use lib)
sqrt	15523	530	0,823%	90%	36%	Re-implement (don't use lib)
mp_GetBuffIndex_u8	73854	992	7,326%	10%	36%	Has been optimised already
d_itod	22765	440	1,002%	40%	20%	Source code not available
cos	2302	562	0,129%	80%	5%	Re-implement (don't use lib)
man_getFieldID_u8			0,740%	10%		Has been optimised already

Fig. 92 Code optimization plan (functions to optimize sorted by priority)

called functions should also be examined to see if they can be converted to inline functions.

It is very helpful to understand the optimizations offered by the compiler being used. Section 8.3.6 will go into this in more detail. The same is true for the special machine instructions provided by the processor, which can often be called using Intrinsic Compiler Functions. Section 8.3.5 discusses the latter in detail.

With mathematical functions, the question arises as to how accurate the result must be. The library functions provided by the compiler for the sine, cosine, root, and other functions are as accurate as possible over the whole range of values, but rarely efficient with regard to the required runtime. With some reductions in accuracy, the runtime requirement can usually be significantly reduced. In the following section, the root function sqrt is examined in more detail as an example.

8.3.2 Optimization of the Function sqrt

The root function sqrt has already played a role in connection with the runtime measurement in Section 5.5. There, the version of the function measured was delivered with the AVR 8-bit toolchain version 3.6.1-1750 (based on the GNU C compiler version 5.4.0). During the measurement the net runtime of the function was between 114 and 646 clock cycles. The overhead caused by the instrumentation is already considered. The timer used for the measurement ran with a prescaler of 1 so that one timer tick corresponded exactly to one clock cycle. The advantage of specifying runtimes in clock cycles instead of seconds is that the specification is independent of the system clock and thus independent of the crystal used.

The optimization of mathematical functions is a topic that has been much researched. The result of this research is that, for the vast majority of mathematical functions, there are alternative implementations that are less accurate but much faster to execute. The easiest way to find such implementations is to search the Internet. The implementation of the square root function shown in the Listing 39 was found through such a search in an application note from the compiler manufacturer IAR [34]. It also appears in other sources.

Listing 39 Efficient sqrt implementation at the expense of accuracy

```
 1  unsigned short sqrtFast(unsigned short x)
 2  {
 3      unsigned short a,b;
 4      b       = x;
 5      a = x = 0x3f;
 6      x       - b/x,
 7      a = x = (x+a)>>1;
 8      x       = b/x;
 9      a = x = (x+a)>>1;
10      x       = b/x;
11      x       = (x+a)>>1;
12      return x;
13  }
```

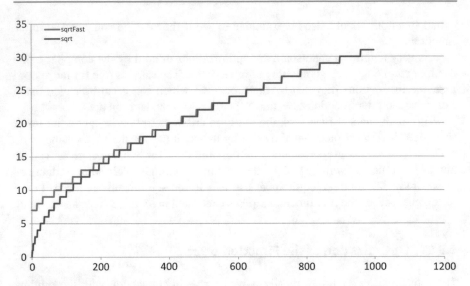

Fig. 93 Comparison of the results from `sqrt` and `sqrtFast` for input values 0 - 999

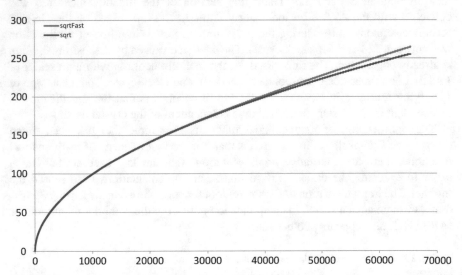

Fig. 94 Comparison of the results from `sqrt` and `sqrtFast` for input values 0 - 65535

With a few additions, binary-shift operations, and divisions, the square root of a number is approximated. The results of the calculation are not as good as the values provided by `sqrt` of the compiler library. Figure 93 shows that a very large error is made for small input values, and Figure 94 shows that too large a result is obtained for larger input values. The higher speed comes at the expense of accuracy. Whether the lack of accuracy causes functional problems is a matter for the function developers to determine. It must also be considered that the calculation includes a

Table 6 Comparison of the
core execution times `sqrt`
vs. `sqrtFast`

	Comparison CETs		
	(all numbers given in CPU cycles for the ATmega processor)		
Function	CET_{min}	CET_{max}	$CET_{\emptyset}$
`sqrt`	114	646	569
`sqrtFast`	464	476	472 (about 17% faster)

division by the input parameter x. If x has a value of zero at runtime, processors that offer appropriate mechanisms will trigger an exception or trap. For these types of processors, the code would have to be supplemented so that it handles the $x = 0$ exception correctly.

Once all these limitations have been considered and found acceptable, the use of the optimized routine can reduce CPU usage.

The runtime savings are shown in Table 6 as an example for the ATmega processor. The runtime measurement presented in Section 5.5 served as the basis for the comparison.

Since the calculation of the square root is now available in the source code, it should also be considered whether it can be implemented as an inline function. This would save another function call and its return. In addition to saving the machine instructions required for the function call, this approach would also be advantageous in terms of a more efficient use of the cache.

The optimization of a single function was discussed in detail as an example. In summary, it can be said that two essential aspects must always be taken into account in every optimization procedure.

Firstly, it must be ensured that the runtime of the optimized variant is actually lower. Many supposed optimizations have resulted in a reduction of source code lines, but turn out to produce inefficient machine code. Here, nothing can replace a measurement or a code simulation.

Secondly, it is essential to check whether the optimized code still delivers the original functionality — and whether it is completely accurate or with an acceptable deviation. In some cases, the effort required to analyze such deviations is so great that the code optimization process must, unfortunately, be abandoned.

8.3.3 Linear Core IDs for the AURIX

In this section, a single very simple function will be examined and optimized. Besides the result of the optimization, the journey to get there plays an important role too. The exact analysis covered here, with the weighing up of advantages and disadvantages of different approaches, and the permanent view of the code actually generated, are classical steps in optimization at the code level. Usually, these steps are supplemented by the runtime measurement but the example considered here is so simple that this step can be omitted on this occasion.

The second-generation Infineon AURIX processor is available with up to six CPUs. Using the Special Function Register at address FE1C it is possible to query the *Core ID* at runtime. However, the IDs are not sequential. For CPUs 0 to 4, this register returns the values 0 to 4, but CPU 5 has the core ID 6. Instead of counting sequentially, i.e. 0, 1, 2, 3, 4, **5**, Infineon counts 0, 1, 2, 3, 4, **6** for the second generation AURIX.

For most applications, it is necessary for the software to provide a *continuous* numbering of the CPUs. Often, this linear ID is used as an index to access an array, and a gap would be a hindrance. So a conversion is needed which returns the entered value if the values 0 to 4 are entered, and a 5 if the value 6 is entered. Since this conversion is required thousands of times per second in many applications, an efficient conversion has a potential for optimization that should not be overlooked.

A simple implementation of the function can be seen in Listing 40. The code supports both the TASKING and HighTec GCC compilers.

Listing 40 Sequential core ID for the AURIX with six cores

```
1  #if defined __GNUC__
2  #    include <machine/intrinsics.h>
3  #endif
4
5  inline unsigned char GetLinearCoreId(void)
6  {
7  #if defined __TASKING__
8      unsigned char id = __mfcr(0xFE1C);
9  #elif defined __GNUC__
10     unsigned char id = __MFCR(0xFE1C);
11 #else
12 #    error "compiler not supported"
13 #endif
14     if (id > 5) {
15         return 5;
16     } else {
17         return id;
18     }
19 }
20
21 int main(void) {
22     return (int)GetLinearCoreId();
23 }
```

The main function is only implemented to call the function. Thus the generated assembler code can be checked quickly. Listing 41 shows the assembler code generated by the HighTec GCC compiler if no optimization is activated. The compiler was called using tricore-gcc.exe -o main.s -S main.c.

Listing 41 Assembler code generated by the HighTec GCC compiler

```
1  GetLinearCoreId:
2      mov.aa   %a14, %SP
3      sub.a    %SP, 8
4      mfcr     %d15, LO:0xFE1C
```

```
 5      st.w      [%a14] -4, %d15
 6      ld.w      %d15, [%a14] -4
 7      st.b      [%a14] -5, %d15
 8      ld.bu     %d15, [%a14] -5
 9      jlt.u     %d15, 6, .L2
10      mov       %d15, 5
11      j         .L3
12  .L2 :
13      ld.bu     %d15, [%a14] -5
14  .L3 :
15      mov       %d2, %d15
16      ret
17
18  main:
19      mov.aa    %a14, %SP
20      call      GetLinearCoreId
21      mov       %d15, %d2
22      mov       %d2, %d15
23      ret
```

First of all, it is noticeable that the compiler implements the function GetLinearCoreId as a regular function despite the inline statement. This in itself is an important finding. Depending on the compiler switch and optimization level, code may not always be compiled as expected. This insight underlines the necessity to check the results by examining the assembler code generated.

In the next step the code is compiled again, but this time with optimization enabled, note the option -O3 in the compiler invocation:

tricore-gcc.exe -O3 -o main.s -S main.c

The resultant assembler code can be seen in Listing 42.

Listing 42 Assembler code with optimization activated

```
1  main:
2      mfcr   %d2, LO:0xFE1C
3      and    %d2, %d2, 255
4      min.u  %d2, %d2, 5
5      ret
```

In contrast to the many assembler commands of the variant without optimization, the optimized version results in just four commands. It makes use of the, quite rare, machine instruction min (minimum) that the TriCore architecture offers. The instruction compares the contents of a data register with a literal (a fixed numerical value) and stores the smaller of the two values in the desired data register. Thus, with a single machine instruction, it is possible to implement what would otherwise have required multiple instructions. In the previous version, the code generated masked out the upper 24 bits since the data type **unsigned char** was specified for the return value of the function GetLinearCoreId.

Strictly speaking, masking out, i.e. setting to zero, of the upper 24 bits is not necessary. A look inside the processor manual of the AURIX reveals that the special function register with the address 0FE1C is a 32-bit register, but only the lower three bits of it indicate the core ID. The upper 29 bits are reserved. Although it is highly unlikely, Infineon could, for example, use bits three to seven (seen from the LSB and starting from zero) for completely different purposes in a future derivative of the AURIX. If masking is already implemented, the upper 29 bits would have to be set to zero explicitly. However, this is not necessary, because the machine command min ensures that the function GetLinearCoreId will never return a value greater than 5. This works even if the register concerned contains a number that is negative. The binary interpretation of a negative number as an **unsigned int** value, as is the case with the min.u machine instruction (the .u stands for 'unsigned'), results in every negative number being interpreted as a number greater than 5 and not as a number smaller than zero.

With the realization that masking is not necessary, the code can be further optimized and implemented as seen in Listing 43. The implementation is now implemented as a macro in order to be completely independent of any compiler options. The compiler can no longer generate function calls for this code. It also uses the datatype **unsigned int**, which is the regular register size of the processor. This eliminates the masking of the upper 24 bits when using the macro if the target is again a 32-bit type.

Listing 43 Efficient code for sequential core IDs for AURIX with six cores

```
1  #if defined __TASKING__
2  #    define GetLinearCoreId( )      __min( __mfcr(0xFE1C), 5 )
3  #elif defined __GNUC__
4  #    define GetLinearCoreId( ) ({ unsigned int coreId_; \
5              __asm( "mfcr %0, 0xFE1C\n\tmin.u %0, %0, 5" : \
6              "=d"(coreId_) ); coreId_; })
7  #else
8  #    error "compiler not supported"
9  #endif
10
11 int main(void){
12     return (int)GetLinearCoreId();
13 }
```

Depending on the compiler, either a 'Compiler Intrinsic Function' (TASKING) or inline assembler (HighTec GCC) is used. The following section deals with Compiler Intrinsic Functions.

Incidentally, Infineon has responded to the demand for linear core IDs and has provided a separate special function register CUS_ID (Customer ID) in newer derivatives to provide them. Using mfcr 0xFE50 you can read a sequential Core ID of these derivatives directly, with the same result as the function GetLinearCoreId, using a single machine instruction.

This example has not only shown that the AURIX offers a special machine instruction that allows the task described at the beginning to be implemented very efficiently, it additionally provided insights into the approach and thinking behind runtime optimization at the code level.

8.3.4 Calculating to Saturation

In many calculations it is a requirement that a result *goes into saturation* rather than overflowing, i.e. it assumes the maximum value it can represent. Similarly, in the case of an underflow, the result should assume the minimum value it can represent.

Listing 44 shows a function that adds two unsigned 16-bit numbers with saturation. If the result were to exceed the value 0xFFFF, i.e. it could not be represented using 16-bits, the result would go into saturation and 0xFFFF would be returned. Many control algorithms depend on the implementation of saturation as, without it, calculations at the range limits of variables could cause dramatic jumps in the results with, sometimes drastic, consequences, such as when positioning powerful robot arms. The primary alternative is to select the value ranges in such a way that overflows and underflows cannot generally occur. There are good reasons, such as code efficiency, to work with smaller ranges and, if necessary, to let the results of calculations go into saturation.

Listing 44 'Typical' addition of two 16-bit numbers with saturation

```
1  unsigned short sadd16(unsigned short a, unsigned short b)
2  {
3      return (a > 0xFFFF - b) ? 0xFFFF : a + b;
4  }
```

Listing 45 shows the assembler code generated by the compiler for an AURIX that implements this function by using a subtraction, an addition, a conditional jump, and further commands, making six in total.

Listing 45 Generated assembler code for Listing 44

```
1  sadd16:
2      mov.u    d15,#65535
3      sub      d0,d15,d5
4      jlt      d0,d4,.L3
5      add      d15,d4,d5
6  .L3:
7      extr.u   d2,d15,#0,#16
8      ret
```

The AURIX provides special machine code instructions for saturation, one of which takes advantage of the implementation shown in Listing 46.

Listing 46 Addition of two 16-bit numbers with saturation using sath

```
1  unsigned short sadd16(unsigned short a, unsigned short b)
2  {
3      return __sathu(a + b);
4  }
```

The resulting assembler code (Listing 47) now only requires three machine code instructions for the same functionality. First, the two 16-bit input variables are added, generating a 32-bit result. This is then 'trimmed' to 16 bits by means of the machine code instruction sat.hu, which implements the saturation.

Listing 47 Assembler code generated from Listing 46

```
1  sadd16:
2      add      d4,d5
3      sat.hu   d2,d4
4      ret
```

It is remarkable that the compiler did not find the optimal implementation on its own despite the highest level of optimization. Only knowledge of the processor-specific instructions by the programmer made possible this significant level of optimization.

8.3.5 Processor-Specific Instructions in General

A large portion of the instructions of one microprocessor can be found in the same or similar form on other microprocessors. Instructions of this type include reading from and writing to memory, copying register contents, additions, subtractions, jump instructions, subroutine calls, etc.

However, many processors offer commands that are less common. The commands `min` and `sat.hu` of the Infineon TriCore architecture used in the previous sections are certainly among them.

Whenever the need arises to insert a specific machine code instruction in the C code, either *inline assembly*, an *Intrinsic Compiler Function*, or *Intrinsic Compiler Macros* can be used. The previous sections have already demonstrated this. The code from Listing 40 uses the Intrinsic Compiler Function `__mfcr` from the TASKING compiler or the Intrinsic Compiler Macro `__MFCR` from the HighTec GCC compiler. MFCR stands for move from core register and it allows the content of special function registers to be read. Intrinsic Compiler Functions and Intrinsic Compiler Macros mostly start — depending on the compiler — with two, occasionally one, underscore.

Listing 43 demonstrates the use of inline assembly (`__asm(...)`). The assembly code contained therein is 'copied' into the generated assembly code by the compiler. The registers used can either be specified concretely, or placeholders can be used so that the compiler is left to determine the register to be used. The exact syntax is compiler-specific.

Since almost all compilers are built on a base that supports a larger number of processor architectures, the special instructions are not always optimally supported. This means that in places where a special instruction would allow a better conversion of the source code into assembly code, the compiler does not necessarily use this instruction. In such cases, the compiler must be specifically instructed to use the special instruction. Section 8.3.4 has shown this using the example of addition with saturation. To be able to use the potential offered by a processor's instruction set, one must be familiar with the instruction set of the processor in question. On the other hand, one must have acquired the necessary knowledge in order to select and benefit from the optimal machine instruction for a specific use case. Knowledge of the instruction set requires diligence and there is no alternative to studying the

Instruction Set Manual of the respective processor. The skillful use of such special commands also requires experience and imagination.

In addition to the machine instructions for special mathematical operations, there are a range of instructions that have no equivalent in the C syntax. For example, the disabling and enabling of interrupts is one of them, which is implemented for the Infineon TriCore with the TASKING compiler using `__enable()` and `__disable()`.

For runtime optimization at the code level, knowledge of the instruction set of the processor used is just as essential as reading the section "Intrinsic Functions" of the compiler manual. The same applies to the chapter of the compiler manual which describes the compiler optimizations. This topic is covered in the following section.

8.3.6 Compiler Optimizations

The compilation of source code to machine code is far from clear-cut. There are a myriad of mapping possibilities, i.e. executables that represent a given source code perfectly correctly. The goal of compiler optimizations is to find the most efficient mapping. In this context, 'efficient' requires a little interpretation. Usually, efficiency is meant in the sense of 'requires little memory' or in the sense of 'requires little runtime'. In the compiler manuals, these optimization goals are described as 'optimize for size' or 'optimize for speed' respectively.

The two goals do not always have to contradict each other. Listing 47 shows the optimized variant of 16-bit addition with saturation, which is both smaller and faster than the initial version.

An example of an optimization that always reduces both memory required and runtime is 'Leaf-Call Optimization'. Here, a function call at the end of a function is replaced by a jump command. Instead of the combination of `call` and **return**, the compiler simply generates a `jump` instruction. The function that is jumped to will itself end with a **return** that then acts as the **return** of the calling function. Or, this called function also ends with an optimized function call, i.e. another `jump` instruction, and so on. The potential for optimization is enormous, especially with a cascade of functions. Listing 48 shows a simple example and Listing 49 shows the corresponding assembler code that the TASKING compiler generates for the AURIX when it is called via `ctc.exe -O2 -o main.src main.c`.

By the way, if `-O3` instead of `-O2` is passed to the compiler as an optimization parameter on the command line, only two of the machine commands of the `main` function remain, namely `mov d2,#0` and `ret`. The compiler recognizes that the functions 'do nothing' and eliminates all calls without replacement. More about optimization parameters will be discussed shortly.

Listing 48 Program for demonstrating leaf-call optimization

```
1  void Function_D(void)
2  {
3      // Do something here...
4  }
5
```

```
6  void Function_C(void)
7  {
8      // Do something here...
9      Function_D();
10 }
11
12 void Function_B(void)
13 {
14     // Do something here...
15     Function_C();
16 }
17
18 void Function_A(void)
19 {
20     // Do something here...
21     Function_B();
22 }
23
24 int main(void)
25 {
26     Function_A();
27     return 0;
28 }
```

Listing 49 Assembler code for Listing 48

```
1  Function_D:
2      ret
3
4  Function_C:
5      j         Function_D
6
7  Function_B:
8      j         Function_C
9
10 Function_A:
11     j         Function_B
12
13 main:
14     call      Function_A
15     mov       d2,#0
16     ret
```

Back to the optimization goals of 'memory efficiency' and 'runtime efficiency'. In many cases the developer has to decide between a small but slower variant or a faster but larger one. The 'loop unrolling' option is a suitable example in this context. In order to reduce the number of jump instructions, loops with a limited number of iterations can be unrolled. The body of the loop then appears in the assembly code as often as the source code requires the loop to be executed.

How can the compiler be made to use one or the other optimization, and how can the optimization goal be specified? There are different ways to do this, and these are listed below.

- Every compiler offers ready-made combinations of optimizations so that the user only has to either specify the rough goal ('optimize for size' or 'optimize for speed') or the degree of optimization. The following switches are commonly used for the degree of optimization: -O0 to disable all optimization measures, and -O1 to -O3 for successively more aggressive optimization. If no optimization options are selected at all, the compiler uses -O0, meaning no optimization at all!

  The optimization options are passed to the compiler when it is called from the command line. If the build environment used supports this, the optimization settings can be adjusted for each source code file if required. Usually, a default setting is made at a central location, such as a Makefile, which is then valid for all source code files.

- For each optimization that the compiler offers there is a separate compiler switch that can be passed to the compiler when called from the command line. The available optimizations are described in the manual and should be studied before undertaking a runtime optimization at the code level.

- Individual optimizations can be activated or deactivated directly in the source code at the level of functions or even within functions. The mechanisms for this are different. The TASKING compiler, for example, provides **#pragma** statements, as the following example shows:

```
1  #pragma optimize u // optimization "Unroll small loops"
2  for (int i=0; i<5; i++) {
3      (...)
4  }
5  #pragma endoptimize
```

The HighTec GCC Compiler — to provide a second example — allows you to enable individual optimizations in the source code for a specific function as follows:

```
1  /* optimize for speed */
2  int __attribute__ ((optimize("Ofast"))) main(void)
3  {
4      return 42;
5  }
```

Debugging software that was compiled with optimization enabled is much more difficult than debugging software that was created without optimization. The relationship between machine code instructions and the corresponding source code becomes increasingly difficult as the number of optimizations increases.

For example, with 'code reordering' the assembler rearranges the sequence of machine code instructions so that the functionality is retained, but accelerates code execution by making optimal use of the pipeline. What benefits the runtime becomes a hindrance when debugging. When the code is executed step-by-step with the debugger, the marker for the currently processed command in the source code jumps back and forth wildly — even without the code containing any jump commands, function calls, or even loops. The flow of the program is barely comprehensible anymore.

So, if code is to be examined for correct functionality in the debugger or tracer, it is usually helpful to deactivate at least some of the optimizations. Some compilers even offer special optimization options that facilitate debugging. For the HighTec GCC compiler, this is the -Og option, which disables those optimizations that are an obstacle to debugging.

Finally, it should be emphasized again how important it is to read the compiler manual in detail to achieve successful runtime optimization at the code level. After each optimization measure, the effect on the generated assembly code must be reviewed and checked by measurement, tracing, or suitable code simulation.

8.3.7   Code Optimization Using the Example memcpy

Some of the optimization approaches presented thus far will be illustrated in this section with a practical example. The approach is very close to that used during the optimization of a real project.

Assume that it was determined by means of runtime measurement or tracing that the function memcpy requires a considerable amount of computing time in the application. This function will now be analyzed and optimized. It is part of the standard C function library and is well known. It copies memory contents from one area to another, and the source code of a simple implementation can be seen in Listing 50.

Listing 50 Simple memcpy implementation for byte-by-byte copying

```
 1  void* memcpy(void* pDest, void const* pSrc, unsigned short n)
 2  {
 3      /* char pointers allows copying with byte granularity */
 4      char* pD = pDest;
 5      char const* pS = pSrc;
 6
 7      while(n--)
 8      {
 9          *pD++ = *pS++;
10      }
11      return pDest;
12  }
```

Parameters pDest is a pointer to the beginning of the target memory area, pSrc is a pointer to the source memory area, and n specifies the number of bytes to be copied. With respect to memory alignment, there are no requirements for the source or the target area. This means that they can also have byte alignment as shown in the right-hand side of Figure 91.

The optimization is carried out using a step by step approach. First, a test function is created that copies a single byte ($n = 1$) and then another of 1024 bytes ($n = 1024$), i.e. one kilobyte (strictly speaking one kibibyte, if the IEC

Table 7 CET for `memcpy` at the various steps described in the text

#	| Memory location			Comp. Opt.	Man. Opt.	Align-ment	CET [ns]		
	`memcpy`	`src`	`dest`				n=1	n=1024	n=1024 per Byte
1	uncached pflash0	uncached pflash0	uncached lmuram	-O0	–	1 Byte	299	123,650	**120.8**
2	cached pflash0	uncached pflash0	uncached lmuram	-O0	–	1 Byte	330	123,552	**120.7**
3	cached pflash0	cached pflash0	uncached lmuram	-O0	–	1 Byte	282	44,128	**43.1**
4	cached pflash0	cached pflash0	cached lmuram	-O0	–	1 Byte	275	44,044	**43.0**
5	cached pflash0	cached pflash0	dspr0	-O0	–	1 Byte	280	43,496	**42.5**
6	cached pflash0	cached pflash0	dspr0	-O3	–	1 Byte	214	17,592	**17.2**
7	cached pflash0	cached pflash0	dspr0	-O3	word	1 Byte	227	17,537	**17.1**
7a	cached pflash0	cached pflash0	dspr0	-O3	word	4 Byte	359	5,830	**5.7**
8	cached pflash0	cached pflash0	dspr0	-O3	Duff	1 Byte	263	14,899	**14.5**
9	uncached pflash0	cached pflash0	dspr0	-O3	Duff	4 Byte	385	24,603	**24.0**

prefixes [4] are used correctly). The execution time required for this is measured using code instrumentation. An Infineon TC275, a first generation AURIX, is used. The processor operates with a frequency of 200 MHz.

Table 7 summarizes the optimization measures for each step and shows the measurement results. The rightmost column indicates the time required to copy a single byte when 1024 bytes are copied.

Each implementation step and its impact are explained below.

Step 1: Initial version without any optimization   First, the code is compiled without any optimization. This means that the compiler compiles the code with the -O0 option and no manual optimization has been undertaken.

The source data `src` is located in PFLASH0, in segment 10 (with address range `0xa...`), that does not, by default have the data cache activated (see section "Contents of the Segments" in the TC27x [35] manual). The `memcpy` function is also assigned to this same memory segment and the program cache for this segment is disabled.

The target data `dest` of the copy process is located in the comparatively slow LMU RAM, i.e. RAM shared by all CPUs of the multi-core processor. Furthermore, the target data within the LMU RAM is located in segment 11 (with address range `0xb...`) for which the data cache is also not active by default.

To review the different memories of the first-generation AURIX, as well as to visualize the access times of the different memories, a look at the extract from the manual in Figure 90 will help.

Neither the source nor the target data have 32-bit alignment.

Step 2: Make use of program cache   The only difference to the software version from step 1 is that the function memcpy has been moved to segment 8 (with address range 0x8...) where the program cache is active.

Surprisingly, the runtime behavior does not change when copying 1024 bytes. How can this be when execution from the cache should be *substantially* faster? Maybe it is because the data accesses are so slow that the execution speed of the code is not the limiting factor. If this assumption is correct, moving the data to faster memory or to segments with an active data cache should improve things. So, before the optimization process is aborted, steps 3, 4, and 5 are focused on the location of the data.

Step 3: Use data cache for source data   Like the function itself, the source data is now moved to segment 8 where the data cache is also active. And, lo and behold, the runtime goes down drastically with the copying process completing around 2.8 times faster.

Step 4: Use data cache for destination data   If the target data is additionally moved to a segment with an active data cache — in this case the segment 9 of the LMU RAM — the net copy time barely changes.

Step 5: Use core-local RAM for destination data   For step 5, the target data is moved to the DSPR (**D**ata **S**cratch**P**ad **R**AM) of the CPU. This RAM is as fast as a cache that does not suffer from cache misses. Thus, the CPU can complete the accesses without any stall cycles.

Admittedly, if you strictly follow the strategy given in Figure 89, a rather large array of 1024 bytes should not actually be moved to the fast DSPR memory.

Interestingly, this change results in very little improvement in the runtime of the code, and this is a very important finding. In this case, a shift to the valuable DPSR does practically nothing. Once again, it is shown that assumptions help to develop new approaches to runtime optimization, but these are worth nothing if their impacts are not backed up by improvements proven by runtime measurement or tracing.

Step 6: Enable compiler optimizations   Admittedly, it is somewhat unrealistic to compile software in a project without any compiler optimizations. However, to illustrate the effect, this example was, until now, configured to compile without them.

With optimizations enabled, the compiler now uses machine instructions that can make a register-indirect memory access that also increases the address register by the number of bytes read after the access. This type of addressing is called 'post-increment addressing' and was introduced in Section 2.3.1. It also minimizes the size of the loop body and uses the loop machine code instruction.

This more compact code becomes even faster by about two and a half times with the highest optimization level -O3.

Step 7: Manual optimization: copying words   As the 8.2.2 section has pointed out, 32-bit architectures struggle to deal efficiently with byte (8-bit) alignment. Copying large amounts of data byte by byte is, therefore, an inappropriate task for an AURIX. In most cases, source and target will also have 32-bit alignment in 32-bit software. Also, the amount of data to be copied will most likely be an integer multiple of 32-bit words.

Using these assumptions, the code can be manually optimized. Listing 51 shows a variant that first checks the assumptions just described. If they are met, the code is efficiently copied word by word, i.e. four bytes at a time.

Listing 51 Manual optimization: word-wise copying (if possible)

```
1  void* memcpy(void* pDest, void const* pSrc, unsigned short n)
2  {
3      unsigned int wordCount = n >> 2u;
4
5      /* Check pDest and pSrc for word alignment. At the same
6         time, check whether n is a multiple of 4 */
7      if(0u==(((unsigned int)pDest|(unsigned int)pSrc|n)&3u))
8      {
9          /* Now use word (1 word = 4 bytes) pointers */
10         unsigned int* pD = (unsigned int*)pDest;
11         unsigned int const* pS = (unsigned int const*)pSrc;
12
13         while(0u != wordCount--)
14         {
15             *pD++ = *pS++; /* Copy 4 bytes at a time */
16         }
17     }
18     else
19     {
20         /* use old implementation of memcpy */
21     }
22     return pDest;
23 }
```

The expression in the `if(...)` query is very compact. It checks whether the lower two bits of all parameters are zero. If this is the case, the two addresses (pDest and pSrc) have 32-bit alignment and n specifies an amount of data to be copied that corresponds to an integer multiple of 32-bit words (4 bytes).

It is not surprising that this measure delivers no improvement whatsoever. The data does not have the right alignment, so in the next step a 32-bit alignment of the data is planned.

Step 7a: Copying of word-aligned data   As soon as source and target data provide 32-bit alignment, the performance increases dramatically. The copy process can now be executed about three times faster.

Step 8: Manual optimization: 'Duff's device'   For the last optimization step, the software developer's box of tricks is used. When Tom Duff optimized animation software for the film industry in the eighties, he developed a C construct that unrolls loops manually that is also transferable to all kinds of use cases [36]. *Duff's Device*, named after him, is a sequence of loop bodies interwoven into a `switch-case` construct and enclosed by an outer, less-frequently executed loop. Listing 52 shows a code fragment that replaces the comment `/*use old implementation of memcpy */` (line 20) in Listing 51. Thus, Duff's Device is only used if the source or target data does not have 32-bit alignment or the number of bytes to be copied is not an integer multiple of four.

Listing 52   Manual optimization: Duff's Device

```
1         /* use Duff's device for non-aligned data */
2         char* pD = pDest;
3         char const* pS = pSrc;
4
5         /* Calculate how many number of iterations are left
6            which are a multiple of four. This is after the
7            first entrance to the switch-case which performs
8            the byte copying which are not a multiple of 4..i.e
9            the trailing bytes. RSH 2 = Divide by 4.
10           Essentially we are loop-unrolling at a depth of
11           4 here.. */
12        unsigned int byteAllignedCount = ( n + 3u ) >> 2u;
13
14 /* surpress warning "no 'break' before case label" */
15 #pragma warning 553
16        switch ( n % 4u )
17        {
18        case 0u:
19            do
20            {
21                *pD++ = *pS++;
22                case 3u: *pD++ = *pS++;
23                case 2u: *pD++ = *pS++;
24                case 1u: *pD++ = *pS++;
25            }
26            /* Pre-decrement byteAllignedCount.
27               This while loop condition is always tested  */
28            while ( 0u < --byteAllignedCount );
29        }
30 #pragma warning restore
```

The `#pragma` on line 15 suppresses a warning indicating that the code closes the `case` blocks without a `break`, indicating a programming error when using `switch-case` in the usual way. Duff's Device is a well secured construct and with `break` commands it would not work anymore, so a temporary disabling of a warning is justified.

The result of step 8 has to be compared with the result of step 7 since both use input data that does not have 32-bit data alignment. It turns out that Duff's Device has a performance advantage of about 15%. If the data has a 32-bit alignment, the code corresponds to the variant from step 7 — with the same result for the runtime.

With step 8 the optimization process is complete. Compared to the initial version from step 1, there is a runtime improvement of about 88% in cases where the alignment and data quantity do not have 32-bit alignment, and a runtime improvement of about 95% if the alignment fits.

Step 9: Program cache back into focus Step 9 is slightly offset in the table to indicate that this step is no longer part of the optimization measures.

When the program cache for the function to be examined was activated with step 2, its impact was surprisingly not noticeable at runtime. The assumption was made that data accesses were the limiting factor. How does the optimized variant behave when the program cache is deactivated again?

For step 9, using the optimized variant of the code, the function was moved back to the segment for which the program cache is *not* active. Now the advantage of the cache is clearly visible, because the runtime increases more than fourfold. For comparison, the results from step 7a have to be used since here, as well as for step 9, 32-bit alignment of the data was ensured.

8.4 Summary and Guidelines for Timing Optimization

As a conclusion for the chapter "Runtime Optimization" a flow chart is presented that can be used for tackling runtime problems. It is particularly suitable for the later project phases where hardware is already available and the main sections of the software are already running.

For space reasons, the flowchart is divided into two figures (see Figures 95 and 96).

Timing problems are very different, as Chapter 6 has already shown. Therefore, this fairly simple flow chart can only provide a rough guide.

The representations in the illustrations are not UML-compliant but are oriented towards practical use. For example, the measures for runtime-optimized memory usage and the optimization measures at the scheduling level can be attacked in parallel, which helps to save time in an emergency. The flow chart is split accordingly.

There is another location where activities can be carried out in parallel. Once the measurement of the access frequencies of all symbols has been completed, the results can be used as a basis for optimization at the code level, while frequently used small symbols can be moved to fast memories. If time is pressing, both of these can be attacked simultaneously.

For optimization at the code level, it is strongly recommended that a code optimization plan be prepared. This was introduced in Section 8.3 and was illustrated in Figure 92.

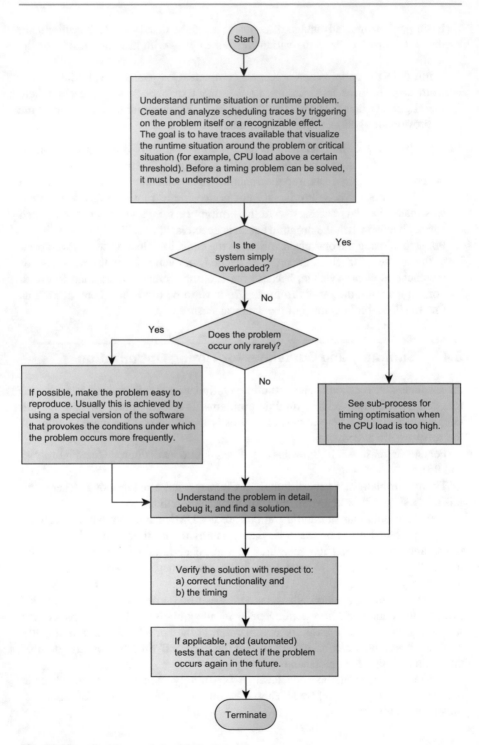

Fig. 95 Flow chart for run-time optimization

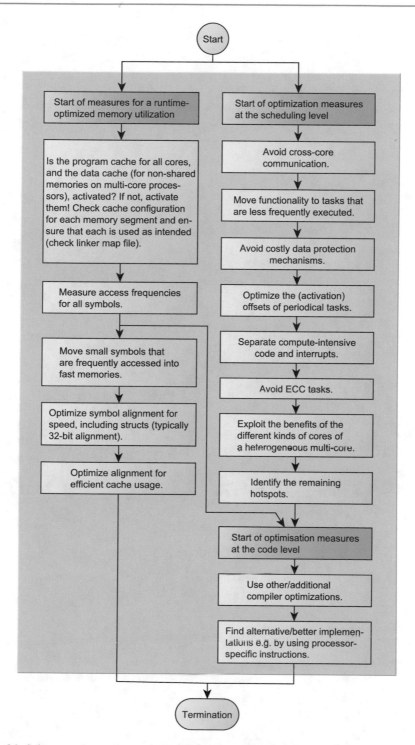

Fig. 96 Subprocess for runtime optimization for cases where overall processor load is too high

Methodology During the Development Process 9

In Section 1.2.1 a brief summary was provided that explained how the topic of timing can be anchored within the development process, and several examples of useful timing analysis measures were provided.

Here, in Section 9 we will expand upon this approach and outline a development process in which timing aspects are considered in all phases of development. The interfaces between each development phase play an important role here that is critical to ensuring the individual phases avoid tackling these issues in isolation, as this can lead to information being created or maintained twice.

9.1 Requirements Related to Timing

If you take a look at the timing requirements for embedded software projects, which are laid down in the requirements specification, you usually don't have to read much. Very often, the specifications are limited to defining the upper limits for CPU load for the various software deliveries over the course of the project. The same applies to the bus load of the communication buses used.

While the software itself has a variety of timing requirements, these are not always explicitly identified, recorded, considered in the design, tested for, or monitored.

This section is intended to provide suggestions on how to approach the issue of timing in the early phase of a project so that subsequent project phases can be built on a solid foundation.

Initially, timing-related requirements should be divided into two categories. The first covers the real timing requirements, those that have the timing of the embedded system in mind, while the second cover those that concern the methods and tools that are related to embedded software timing. This second category does not include timing requirements at all, but is still often classified under the heading 'timing requirements'.

© The Author(s), under exclusive license to Springer Nature Switzerland AG 2021
P. Gliwa, *Embedded Software Timing*,
https://doi.org/10.1007/978-3-030-64144-3_9

9.1.1   Timing Requirements

To get closer to the topic, we will first provide a few examples of concrete timing requirements.

Startup time   It wasn't long ago that you would regularly observe many developers getting their morning coffee, after having turned on their PC, arguing that the PC had to boot anyway. Since the introduction of SSDs and fastboot have become more widespread, other excuses now have to be provided to justify early coffee enjoyment.

The booting of a PC corresponds to the startup time for embedded systems. It is the time that elapses after a reset until a certain functionality is available. For most embedded systems, the startup time is usually in the low millisecond range. For many automotive ECUs, there is a requirement that they must be able to show presence on the bus after 100 ms, i.e. that they must be able to send and receive network management messages.

But there are also counter-examples, that is embedded systems, which do not start up as quickly. The regional trains to Munich coming from Kochel or Mittenwald are coupled together in the town of Tutzing in order to continue as a single train. As a passenger, the coupling process is perceived as a physical jolt. Afterwards, however, a few minutes pass before the train finally starts moving because the software is restarted after the coupling process. Obviously, the issue of timing was neglected during development.

End-to-end timing requirements   Here, 'end' means the two ends of a chain of events. This can be a sensor (for example a brake pedal) on one side and an actuator (for example the power electronics for controlling the brake lights) on the other. An end-to-end timing requirement would be that the brake lights illuminate latest 200 ms after the brake pedal is pressed.

Maximum allowed net runtime (CET)   In Sections 5.8 and 5.9 it was shown how scheduling can be optimized and verified using these techniques. The prerequisites for this are the maximum net runtimes of tasks and interrupts. Therefore, the timing requirements with respect to the maximum net runtimes have to be available, latest, when these approaches are applied.

However, even when neither scheduling simulation nor static scheduling analysis are used to their full extent, the specifications with regard to maximum net runtime are still useful. They can be understood as a more finely resolved requirement regarding the maximum permitted CPU utilization. It is true that, at an early stage, there will probably be only very vague ideas about how much runtime is allocated to the individual tasks and interrupts. But a vague idea is better than no idea. The specifications should allow for the adjustment of the required values for the maximum net runtimes over the course of the project so that they reflect the current conditions. Section 9.1.2 looks at this concept in more detail.

Periodicity   Most embedded systems, especially those that contain one or more control algorithms, contain code that must be executed periodically with a certain time interval. If the actual time interval, the Delta Time (DT), deviates too much from the required value, the control algorithm no longer functions correctly. The controller might become unstable and start to oscillate, with potentially catastrophic consequences for any connected mechanical systems.

In the software design of periodic portions of code, the question that must be answered, in addition to the question of the desired delta time, is: What maximum deviation from this desired value is allowed? In the practical example described in Section 6.7, the customer had formulated a requirement for the periodicity of the network management messages in the specifications in an exemplary manner: "The messages must be present on the bus every 10 ms; a deviation of plus/minus 1 ms is permissible".

The fact that this requirement was not subsequently verified in the case in question, and was in fact breached, is a different matter.

Execution order   The requirement regarding the order of code sections does not appear to be a timing parameter at first glance. After all, a time specification in seconds (or milli- or microseconds) is missing. Nevertheless, the order is very important information for the system architect or the operating system configurator. If the system architect knows the dependencies that exist between the execution of the runnables, they are able to relocate the runnables of a multi-core system between the cores, or between the tasks of a CPU, to achieve an improved system utilization.

While the order in which the runnables are processed is only one aspect of this work—the bindings are another important one—this type of optimization is not possible without order requirements.

Maximum allowed response time (deadline)   Scheduling simulation and static scheduling analysis were already mentioned above in connection with the maximum allowed net runtime, the WCET. A central output variable of these techniques is the response time of tasks and interrupts. The response time describes the time span between the need for an execution (activation) and the completion of the execution (termination or end).

Whenever a timing requirement can be derived from a closer look at a certain functionality, it should be formulated accordingly in the specifications. In some cases, a deadline can be derived from the individual sections of an event chain.

Maximum allowed age of data   The age requirement for data is orthogonal to the response time requirement defined for tasks and interrupts.

Here too, sections of a chain of events can be represented as the age of data. Furthermore, the practical example described in Section 6.4 can be described as needing requirements for the age of an item of data.

Maximum allowed CPU load   Finally, we should mention again the CPU load mentioned above. It is often a thorn in the side of academic timing experts. With AUTOSAR TIMEX it cannot even be formulated. However, since it has become a habit of many developers and managers in the past to include this quantity in their specifications and refer to it in the course of project development, it has earned the right to exist in future specifications. Further reasons for why CPU-load is a reasonable requirement were described in Section 4.3.

9.1.1.1 Identification of Timing Requirements Through Interviews

So, how do you approach the definition of timing requirements for an embedded system? One approach is to interview the function developers, network experts, and integrators involved—in that order. The aim is to elicit the actual requirements from the interviewee, even if he or she may not be aware of the existence of these requirements at first. Chris Rupp [37] has dealt with this discipline in detail.

Every function developer is asked which timing requirements exist for the functionality they develop. Depending on the level of abstraction at which development takes place, the questions relate to functionality at the physical level, the model level, the software component level, the runnable level (if the project is based on AUTOSAR), the C function level, or even lower.

The integrators usually have a lot of experience with embedded software timing and can usually contribute requirements that the function developers did not have in mind.

The questionnaire shown in Figures 97 and 98 can be seen as a recommendation or starting point for your own questionnaire. A questionnaire is filled out for each timing request. The collected timing requirements must be included in the requirements specification.

The questionnaire is available as a Word document in the online resource accompanying this book.

9.1.1.2 Formats for Timing Requirements

For AUTOSAR projects, the insertion of timing requirements into the requirements specification can be done formally using AUTOSAR Timing Extensions (TIMEX). Section 10.3 covers this topic in more detail and Section 6.8 has highlighted a series project that has successfully followed this path. However, it should also be mentioned here that, at least as of summer 2020, the use of TIMEX for the formulation of timing requirements is the absolute exception. There is simply too little support provided by the relevant tools. Nobody is going to sit down and write the timing requirements in AUTOSAR XML as specified in the TIMEX standard. The AUTOSAR XML syntax was designed to be generated, not manually created. Additionally, the semantics of TIMEX are very complex. Overall, they are very flexible, comprehensive, and logical, but require a considerable training effort in advance before they can be used with confidence.

For non-AUTOSAR projects, and as an alternative to TIMEX or as an interim solution on the way to TIMEX, the timing requirements found through the interviews can simply be recorded in writing. Every developer understands the meaning of the maximum allowable CET for a function, runnable, or task. The upper limit of jitter for a control algorithm is just as easily understood. Specifying maximum or minimum allowed values for selected timing parameters has several advantages.

Questionnaire regarding timing requirements

Date: _____

Project name/ID: _____

Filled out by (name): _____

Brief description of the timing requirement:

_____

_____

_____

What is the requirement related to?

Description: _____

Please mark with a cross:

- ☐ Code/functionality
    - ☐ Physical level
    - ☐ Model level
    - ☐ Software component
    - ☐ Runnable
    - ☐ C Function
    - ☐ Code level
    - ☐ Other: _____
- ☐ Data
- ☐ Other: _____

Figure 97 Page 1 of the questionnaire on timing requirements

- Permissible minimum and maximum values for selected timing parameters are easy to understand.
- Permissible minimum and maximum values for selected timing parameters can be formally specified using either a self-defined format, a format used by a timing

Which timing parameter is relevant?

Please mark with a cross:

- ☐ CET   (Core Execution Time)
- ☐ DT   (Delta Time)
- ☐ RT   (Response Time)
- ☐ GET   (Gross Execution Time)
- ☐ ST   (Slack Time)
- ☐ NST   (Net Slack Time)
- ☐ IPT   (Initial Pending Time)
- ☐ JIT   (Jitter)
- ☐ PRE   (Preemption Time)
- ☐ Execution order (e.g. related to runnables)
- ☐ Data age (time between read and write or between send and receive/usage of the received data)
- ☐ Boot/init: from _____ to _____
- ☐ Not applicable
- ☐ Other: _____

What type of value is it?

Please mark with a cross:

- ☐ Minimal allowed value
- ☐ Maximal allowed value
- ☐ Average value
- ☐ Not applicable
- ☐ Other: _____

What is the value (e.g. in μs or ms)?

Limit or required value: _____

    or

Execution order: _____

Figure 98 Page 2 of the questionnaire on timing requirements

tool, or a standardized format. ASAM ARTI is a potential option here, described in more detail in Section 10.4.2.

- Permissible minimum and maximum values for selected timing parameters can be processed directly by timing tools.

Various timing analysis tools can be used to check the timing requirements. These can be scheduling simulation, static scheduling analysis, timing measurement, or tracing. All of these techniques, described in more detail in Chapter 5, provide timing parameters as outputs that can be compared to the timing requirements, answering: Are the timing parameters, as determined by the timing analysis tool, within the specified limits? Such checks are as simple as they are pragmatic.

Depending on the project phase, this is implemented as a fully automated system as part of the build process or automated testing. Experience has shown, time and again, that projects that install such automated tests are rarely surprised later on by timing problems. Section 9.6 goes into more detail on this topic, but it should be noted that the time savings that automated timing tests can deliver can be enormous. Once set up, they require very little maintenance. The execution of the tests is done using scripts, so the developers don't even need to lift a finger.

As soon as automated timing tests result in the early detection of even a single timing problem, they have already paid for themselves.

9.1.2   Requirements Regarding Methodology and Tools

While the previous Section 9.1.1 dealt with concrete and usually very project-specific timing requirements, we will now look at the requirements related to the methods and tools for timing.

First of all, it must be considered in which project phases the timing analysis should play a role and to what extent. The following sections, as well as Chapter 5, can and should provide recommendations for this. For safety-relevant projects there are also concrete requirements regarding timing analysis. Section 11.2 deals with this aspect in more detail.

A quite obvious requirement for the tools and methods is to capture and verify the previously specified concrete timing requirements. So, which tools are available for the collection and verification? The CETs can be determined and monitored by means of timing measurement, tracing, code simulation, and/or static code analysis. For timing parameters at the scheduling level (e.g. response time, delta time, or CPU load) timing measurement, tracing, scheduling simulation, and/or static scheduling analysis can be used. Simulation and model-based approaches can be used early on, even before hardware and software are available, and allow for the rapid evaluation of different configurations in short evaluation cycles. Measurement and tracing allow insight into the real system independent of any errors or shortcomings the model or simulation may have. It is especially important to have a look at the real system when it behaves differently when compared to expectations or the model.

An essential aspect of this book is to convey the basics for making optimal decisions when choosing tools and methods. There is no simple rule available for this selection as the requirements are too often too developer- and project-specific. Of course, the currently used landscape of tools and existing internal development procedures will also play an important role in this decision.

Once the considerations regarding which analysis techniques to use have been completed, the question arises as to which project partner is involved in which form. This defines which requirements, methods, and tools need to be included in the specifications. A concrete example illustrates this below.

The manufacturer of a flight controller in the aviation sector purchases essential parts of their software, including the basic software, from a supplier. The supplier also undertakes the software integration. The manufacturer also develops large sections of the application software. A timing measurement technique independent of the hardware is to be used for timing optimization and timing verification. The manufacturer has specified this in the specifications. What was forgotten, however, is the requirement that the measurement technology must also be available to the manufacturer for the development of *his* sections of the software. As defined in the requirements, the supplier uses measurement technology and the results are documented and made available to the manufacturer. The measurement technology used is a tool developed by the supplier that can and may only be used internally. The manufacturer now finds themselves in the shocking situation of not having any measuring technology available for the development of their portion of the code. It is also extremely difficult to install a solution of one's own, because the desired type of measurement technology requires integration into the basic software.

Usually, the timing analysis tool to be used by suppliers is not prescribed. Sometimes, however, prescribing such a tool is useful and can be implemented by means of an inclusion within the requirement specifications. This is already the case when it comes to the compiler to be used. If no concrete tool is to be defined in the specifications, the desired features must be formulated with sufficient precision as individual requirements.

Finally, the following collection of questions, resulting from day-to-day experience, should help when you are defining the specifications and need to consider what the most important aspects regarding tools and methods related to timing are. It should be noted that not all questions require answers for every project!

- Is it a safety-critical project? If yes, have the requirements of the relevant safety standard regarding timing been taken into account? Chapter 11 deals with the topic of safety in more detail.
- Can the intended timing analysis technique adequately capture and test/verify all concrete timing requirements?
- Is there also an analysis technique that can be used for efficient timing debugging (visualization of the real system at the scheduling level)?
- Should analysis and optimization of scheduling be possible without hardware?
- Which project partners should have access to which tools?
- Which project partners should have access to the results of the timing analysis?
- Where and at what times should which tests take place and to what extent? For the 'where' the following can be considered:

- using MIL (model in the loop)
- using SIL (software in the loop)
- in the laboratory
- with PIL (processor in the loop)
- at the HIL (hardware in the loop)
- on the subsystem at a test station
- in the target environment (in the plant, in the vehicle, in the aircraft, etc.)
- Is it clearly defined how CPU load is to be determined? Only if the definition/-configuration is the same can comparable results be generated (see Section 4.3).

9.1.3   Generic Requirement Templates

The series project described in Section 6.8 was not only a successful and forward-looking 'timing project'. Many requirements regarding timing analysis were generalized after the creation of the project's requirement documentation and collected into a pool for use in the requirements documents of future projects. Over the years, this pool of *timing requirements text templates* has been maintained, refined, and supplemented more and more extensively and in discussion and agreement with various timing experts.

Whenever a new project starts, this pool can be drawn upon to create the specifications. The time saved when creating the specifications is significant and offers a major advantage. What is even more important is that, when collecting the requirements, it is ensured that the authors of the specifications have at least considered the timing aspects that are important for the successful development of a safe ECU.

Such approaches are increasingly gaining recognition and dissemination due to their significant advantages. And, it can also be observed that the number of timing problems in projects using these timing requirement text templates is drastically lower than in projects where the issue of timing is approached from scratch, or even ignored completely when the specifications are drawn up.

However, experience has shown that, as good as this pool of timing requirement text templates are, they need to be maintained and communicated. Maintenance here means continuous updating with respect to the constantly changing development processes and development environments, as well as the transformation of the tool landscape. Even more important, however, is the training of those responsible for writing requirements specifications. In the first step, they must be aware that a pool of requirements document text templates exist. If they also have a solid basic knowledge of timing, then everything has been done correctly.

9.2   Collaboration During the Development

The term 'project partner' is used here to mean the relationship between a client and their suppliers. As far as timing is concerned, there is not much to consider when setting up the cooperation of project partners. The core issues have been identified with the requirements described in the previous section. Nevertheless, a lot of time can be saved if additional topics concerning the cooperation are discussed and specified early on.

An important aspect of this is the exchange of timing related information without disclosing intellectual property (IP) where this is not desired. Figure 99 shows a typical scenario of collaboration between the customer and the supplier when both partners contribute software elements to the project and both want to protect their intellectual property. To ensure this, code is exchanged at the object code level rather than at the source code level.

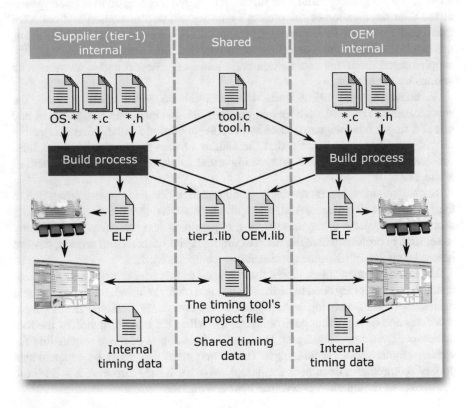

Figure 99 Joint project development with IP protection

All non-IP-relevant information required for timing analysis should be located in a common area accessible to all project partners. This information includes, for example, the project files of the timing analysis tools regardless of whether they are scheduling simulation, static scheduling analysis, tracing, or any other appropriate technique.

Whenever instrumentation is involved—this is the case with software-based tracing, software-based time-of-flight measurement, and hybrid software/hardware tracing approaches—the generic part of the timing tool's source code also belongs in the common domain. This includes those modules that contain the configuration, such as the size of the trace memory in software-based tracing, if the trace memory is in the RAM of the processor.

Each project partner is also free to instrument its protected parts as well to create detailed measurements or traces that will not be made available to everybody, independently of the common area. It should be noted that, as a rule, an identification (ID) will have to be assigned for each instrumentation point. In order to avoid that IDs are used twice, it makes sense to assign an ID range to each project partner at the beginning of the project. If there are two partners, it is advisable that one of them is assigned the lower range and the other the upper one. For example, if the tool supports user-defined events that are identified by an 8-bit ID, the supplier could be assigned the IDs 0x00 to 0x7F and the OEM the IDs 0x80 to 0xFF. Without further agreement, each of the partners can now create user-defined events during the course of the project without any collisions. If there are several partners, several areas must be created.

A cooperation prepared in this manner enables all partners to carry out timing analysis—both for the entire system and their own code—flexibly, in detail, and without complications.

9.3   Timing Concept, Scheduling Layout, and OS Configuration

Once the concrete timing requirements have been fixed and the processor selection has been made, the timing concept can be approached. From there the scheduling layout is then derived and, finally, the operating system configuration results. In a perfect world, these steps would take place before the processor selection and would have a significant impact on it. A well-founded load estimate can only be made when there is an approximate picture of which software elements will run on which processing cores.

However, the performance of a processor is only one of the decision criteria in the selection process and, therefore, the processor is usually already known before the timing concept is approached.

A pleasant exception, and a very structured approach, is shown in the practical example at the end of this chapter, described in Section 9.9.

Unfortunately, there is no simple rule for the distribution of the software to the different cores of the processor(s). Each project has very specific requirements and the aspects that play a role in their distribution are sometimes contradictory. Perhaps at some point there will be tools that can do the distribution themselves. Until then, experience and expertise is needed—knowledge which this book tries to impart.

Section 8.1 has already listed some aspects that can be taken into account when designing a system in order to improve its performance. In addition to functional aspects and timing, safety also plays a role, answering questions such as: What requirements result from the safety concept? Do memory areas have to be separated? Does it make sense to locate safety relevant code on one or more dedicated cores?

As soon as the initial ideas for a timing concept and scheduling exist, they can be transferred to a scheduling simulation. The ideas thus become tangible and can be refined as more information is added. The simulation also helps enormously to further develop and sharpen the ideas. What works, what doesn't, and what effects a particular approach has on the scheduling can be quickly identified in the simulation.

Whether a concept really does meet the requirements in all situations can then be verified using static scheduling analysis.

9.4   Timing Debugging

The word 'bug' has become a synonym for software errors. The 'debugger' is therefore a tool to eliminate software bugs. In most cases, the functional aspects are the focus of attention. By stopping the software at a point where an error becomes noticeable, the contents of variables can be checked and, in single-step mode, the software's flow can be traced.

For debugging timing problems, however, this approach is only of limited help. Many embedded systems are so embedded and dependent on their environment that it is not possible to stop and then execute the software step by step. Even if this were the case, classical debugging is not the method of choice for most timing problems. It focuses on the code level and most timing problems are found at the scheduling level. None of the timing problems described in Section 6 could be solved by classic debugging.

Insight into the scheduling level of the real system is best provided by tracing. It is irrelevant whether the scheduling traces were created using instrumentation or hardware. What is important is that the execution of interrupts and tasks, or threads and processes, on all relevant CPUs, as well as the exchange of relevant data, can be visualized.

9.5   Timing Optimization

It is unusual to regard the timing optimization as a fixed step in the development process. In most projects, timing debugging and subsequent timing optimization is only started when timing problems occur.

It is certainly worthwhile, however, to consider capacities for optimization during project planning. In the best case you will find that everything is in order when it comes to timing optimization and, thus, optimization is not required. If this is not the case, please refer to Section 8 which deals with the topic in detail and shows a whole range of possible optimization approaches.

9.6   Timing Verification

Timing verification stands and falls with the availability of automated timing tests. If they are available, and the automated tests are executed nightly, the probability of being surprised by timing problems is extremely low.

More and more projects are going one step further. They run the timing tests every time they commit or push their code—that is, when a change is committed to version control. This approach, known as 'Continuous Integration', has already become established for functional testing.

Automated tests are implemented with different environments. Unit tests usually run on an $\times 86$, i.e. locally on the PC or a test server. Section 5.4.2 discussed in more detail how unit tests can be extended relatively easily so that, in addition to functional testing, they can rudimentarily capture and monitor the CETs of the functions under test. This allows profiling at the code level without significantly extending the duration of unit tests.

The idea of running timing tests in parallel to functional tests can also be transposed to higher levels. System tests are performed on the HILs and at the subsystem test stations and, again, profiling can be run simultaneously. In addition to the CETs, the acquisition and monitoring of the timing parameters of the scheduling level—especially CPU load, RT (Response Time), DT (Delta Time) and NST (Net Slack Time)—is possible and useful.

The timing tests offered by a HIL can be divided into the categories described in the following sections. If possible, each category should be assigned its own test phase when performing the automated tests.

9.6.1   Test Phase "Profiling"

Repeated briefly, profiling describes the process of determining timing parameters. This can be done directly using timing measurements or indirectly using tracing. Section 5.2.2 has explained this in detail in Figure 40.

The individual functional tests target specific functions and operating states. In most cases it will be useful to record a separate set of profiling data for each of these tests and store it together with the functional test results. Some of the tests will certainly include timing parameters specific to the test. If, for example, all diagnostic jobs in a test section are checked individually, at least the CET of the diagnostics handler and the total duration of each individual job from request to implementation should be measured.

9.6.2   Test Phase 'POI (Point of Interest) Tracing'

Often, embedded systems are designed to be in a normal state most of the time. In this state, a controller can undertake its work, route its messages, wait for input, etc.

Then there are events that cause this normal state to be left behind, which often have a more or less significant impact on scheduling. From the point of view of timing analysis, such events are 'POIs (Points Of Interest)', because it must also be ensured that the scheduling and timing here remains in order during this deviation from the normal state.

Events of this type include error handling, the execution of diagnostic jobs, the execution of additional functionality, or a change to another operating state. Section 6.7 described such a change as the cause of a timing problem. The problem was resolved and subsequently Point Of Interest Tracing was used to ensure that the problem will not occur again in the future.

One of the objectives of unit tests is to test exceptional conditions in a targeted manner, i.e. to include tests that go beyond those test vectors that cover normal operation.

This philosophy also underlies Point Of Interest Tracing. The POIs are specifically brought about in automated tests and traces are created that show their scheduling during the POI. If something unusual or unexpected occurs, this is immediately noticed either by examining the trace or by (automated) checking of previously formulated timing requirements. If everything is considered to be in order, the trace is stored together with other test results and the error-free scheduling at the POI is thus documented.

9.6.3   Test Phase 'Corner Cases'

In the previous section parallels to unit testing were shown. Such parallels also exist for the 'corner cases' (rarely occurring edge cases). In unit testing, a specific attempt is made to cover them by sampling the marginal areas of the input variables.

Unfortunately, this is not so easy to transfer to scheduling. Even if one knew exactly which combination of tasks and interrupts together with which CETs (or other timing parameters) lead to the 'worst case', such a situation could not be provoked easily.

For the analysis of the corner cases, therefore, analysis techniques are used that are independent of the real hardware and environment, such as scheduling simulation and static scheduling analysis.

9.6.4   Test Phase 'Empirical Determination of Headroom'

The goal of this test phase is to find out how much headroom the software currently has before it runs into a timing problem. It could also be called 'robustness analysis'. The approach is as pragmatic as it is simple. A delay that is scalable at runtime is built into the code for which the analysis is to be performed. This delay should be adjustable to consume a defined CET. While the relevant tests are now repeatedly executed, the CET of this delay function is slowly increased. Meanwhile, all timing requirements are checked, especially those concerning DT, RT, and CPU load. As soon as a timing request is violated, the current CET of the delay is recorded as the test result for the affected code section and the tests performed.

Of course, the resulting value is not to be regarded as the CET by which the affected code location can be extended. The 'available time' is, of course, below this. Nevertheless, this empirical approach is very helpful for obtaining a feeling for where the project currently stands. A very small value of the CET for the delay should be interpreted as an alarm signal. In such cases there is an acute need for action.

The NST (Net Slack Time) was covered in Section 4.1.1 and is closely related to the empirical determination of available time. Instead of using a delay function to use up the available time, the runtime situation is viewed from the other perspective by calculating the available free time.

9.7 Early Consideration for Future Functionality

The delay functions introduced in the previous section can also be used in other ways. During the course of the project, placeholders for future functionality can be added. Although these cannot represent the functionality itself, they can represent the runtime required for the functionality. Figure 100 illustrates this. The project history with its software releases can be seen on the time axis. Instead of 'releases', the term 'integration stages' is often used. The y-axis shows the features, the functionality, and the CPU load. There is a certain correlation here as additional functionality demands additional computing time and, thus, increases the CPU load.

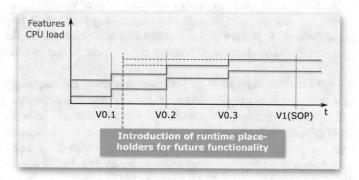

Figure 100 Safeguarding of future software versions with regard to timing

So what is the idea behind early consideration of future functionality? The figure above provides an example. After release V0.1 delay functions, placeholders for future functionality, will be implemented. The project plan is used to look into the future and consider the impact of features that are planned for the next release. Perhaps the upcoming extended functionality for release V0.2 includes a feature that has already been used in other projects and has been measured with regard to its timing requirements. Thus, a placeholder with a CET of 620 μs can be added to the 10 ms task, and another one with a CET of 81 μs can be created in the 1 ms task.

A look at the features of the V0.3 release reveals that a new development is pending for integration that has not yet been used in any other project. Therefore, no runtimes can be imported from other projects. However, measurements may have been made on a PIL (PIL = Processor In the Loop), possibly in connection with unit tests as explained in Section 5.5.2. CETs determined in this manner can then be incorporated into the software via appropriate placeholders long before the actual functionality is integrated for the first time.

Analogous to this approach, future communication on the buses can also be estimated. If, for example, a CAN bus is used, planned messages can be sent at an early stage in their expected transmission patterns so that the bus load is provisionally defined. This extended system also starts to reflect much more closely its future form with regard to delays to lower priority CAN messages.

With this modified platform extensive tests can now be carried out. The functionality corresponds to release V0.1 but the timing is the same as expected for release V0.3. Such tests are not limited to the laboratory or HIL; they can also be extended to the final environment of the embedded system, for example the vehicle in the case of an automotive ECU. The future timing behavior can be experienced in the truest sense of the word.

It is possible that, after the insertion of all the placeholders, it will turn out that the system is no longer able to run at all. What looks like a disappointment at first glance is, in reality, an exceptionally valuable insight. It is made clear early on that, without timing optimization, version V0.3 of the software will not be stable. Without the placeholders, this would probably only have been noticed when integrating V0.3 and at a time in the project when the stress levels are very high due to the pressure of adding new features. With each problem that arises in a development phase it can be very time-consuming to determine whether the problem is due to functional or timing issues.

This is a very decisive advantage of the approach using the mapping of future functionality with (runtime) placeholders: it spreads possible problems over the course of the project. Timing problems can be investigated and solved early on, independently of specific functionality, while the addition of new functionality typically impacts the timing only imperceptibly. The time at which the placeholders are activated can be freely selected. A comparatively quiet phase after the completion of a release—V0.1 in the example—is recommended.

9.8   Timing Supervision in the Final Product

Verification of timing is usually seen as part of the development of an embedded software project. But timing analysis can also be useful afterwards, during the regular use of the system.

In a simple form this is undertaken with a *watchdog*. A hardware timer counts down and triggers a software reset when it reaches zero. In a low priority task—for example, the background task—the software assigns the counter its initial value. If the software is working as intended, the counter never reaches zero because the background task gets called frequently enough.

However, if the software 'hangs' due to an error, the counter is not reset and the watchdog counts down to zero causing a software reset to be triggered. The software should include the ability to determine the cause of the last reset as part of its initialization. Any unwanted reset, such as a watchdog reset, should result in an entry in the diagnostic error buffer so that it can be determined that there is a problem.

The watchdog provides little information for error analysis. The diagnostic error buffer entry merely indicates *that* the watchdog triggered, but not *why*.

Conversely, the absence of watchdog resets does not mean that the system is free of timing problems. The watchdog is just a final backstop.

Monitoring of the timing can be much more granular if regular timing measurement (see Section 5.5) is used as its basis. This allows you to determine and monitor specific timing parameters at runtime. Minimum and maximum values can be stored in non-volatile memories, such as NVRAM or EEPROM, and read out later when the embedded system is accessed.

In such a scenario, the code that represents the timing measurement is no longer just a pure measurement technique, it becomes part of the software. It must therefore also meet the quality and safety requirements of the production software. For use in a safety-relevant automotive control unit, such as a steering or braking system, certification according to ISO 26262 ASIL-D is advisable or even required.

Even with the timing measurement, the analysis of error causes will be difficult in most cases. Tracing is the method of choice here to understand, and then solve, timing problems. If the non-volatile memory has sufficient space, traces can be stored persistently in the event that a timing violation is observed. Memory is getting consistently larger, which means that it is increasingly possible to store traces that are short in absolute terms, but usually sufficiently large for analysis. In many cases, the larger problem will be collecting traces from the field. In some cases, such as in industrial plants, remote access is available and traces can be retrieved using this capability.

9.9   Positive Example: CoReMa by Vitesco Technologies

As a conclusion for the Chapter 9 a very successful approach by a supplier in the automotive sector will be presented. It is no coincidence that the protagonist of this success story comes from the powertrain sector. Engine control units were the first control units ever in the vehicle and the departments concerned have acquired a corresponding amount of experience. This applies to Robert Bosch just as much as it does to Vitesco Technologies (formerly the power-train division of Continental), DENSO, or any of the other long-standing players in this market.

For its engine control units, Vitesco Technologies has been systematically collecting information about the resources required by the software for years and over many software generations. This includes the RAM, stack, and flash requirements, as well as the timing requirements. The figures regarding RAM and flash requirements can be easily read out from the linker map. Static code analysis is used for the stack requirements, and the CETs are determined by means of software-based measurements embedded in automated tests on several HILs.

All numbers are determined at least at the level of what AUTOSAR would call runnables, that is, for the functions that are called directly by the tasks. A certain set of functions represents a certain functionality—again analogous to the software components (SW-Cs) in AUTOSAR, which are composed of a certain set of runnables.

Most software components are developed independently of a specific project and can be used again and again for different projects, similar to the building blocks in a construction kit. The processor types and compilers used in these projects are not always the same but, without fail, the resource information is always systematically collected in a database. The methodology and the database, together with the associated tools, is named CoReMa (**Co**re **Re**source **Ma**nagement). Over time, this has given Vitesco Technologies a very detailed picture of which software component uses how many resources and under what conditions (see Figure 101). Each column "Env x" contains the resource data for a defined environment (**Env**ironment). The parameters that make up an environment include the processor used, the clock frequency, the compiler used, the compiler switches, the operating status (including the motor speed), etc.

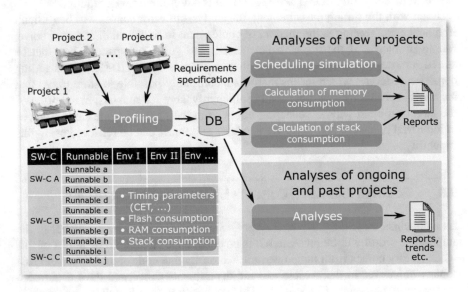

Figure 101 CoReMa workflow and database structure

This alone is very valuable for software development and safety. For example, new values resulting from software changes are automatically compared with the previous values. If a software change unintentionally changes the resource requirements by leaps and bounds, this is detected immediately and not just weeks after the integration of the component when, suddenly, sporadic errors occur in the field that then have to be analyzed in a time-consuming and costly manner.

For future projects, CoReMa opens up even more possibilities. If Vitesco Technologies receive the specifications for an upcoming project from a customer, the developers can virtually assemble the future control unit by means of scheduling simulation and based on the CoReMa data. The requirements specification defines the desired functionality. The required building blocks (software components) can now be selected from their database and integrated into a simulation that represents the desired functionality.

Using this scheduling simulation, the timing design of the software can be set up, the expected CPU load can be estimated, and so on. This allows early answers to crucial questions such as: Is the cheaper processor A sufficient or does it have to be processor B? The processor change is quickly accomplished in the simulation. The database offers the net runtimes of the software components for different processors.

Vitesco Technologies is therefore able to provide a reliable estimate of the future scheduling at a very early stage. As soon as the software is running on the early samples of the new hardware, efficient tracing is available and the traces generated allow for comparison with the simulation. In addition, very efficient (timing) debugging is possible—either on the developer's desk, at the HIL, or in the vehicle.

This approach is a blueprint for anyone who wants to keep an eye on, and control, timing throughout the course of a project. Having far fewer timing issues to face is a great pay off after the initial effort of introducing such an approach. Vitesco Technologies has demonstrated this—their developers rarely have to deal with unforeseen timing problems.

9.10 Summary

How can timing be considered at each individual phase of (embedded) software development? Chapter 9 has tried to answer this question, or at least provide some food for thought. The goal of considering timing is to develop safe and highly reliable embedded systems at low cost.

The recommended measures range from the determination of concrete timing requirements and timing analysis requirements, to dedicated timing tools and the supervision of timing in the final product.

A short description of the process at a large supplier to the automotive industry rounded off Chapter 9. Over many years, the CoReMa database at Vitesco Technologies has proven itself many times over. It has allowed Vitesco Technologies to keep an eye on timing by means of automated processes, and even to look a little into the future with new projects by means of scheduling simulation.

AUTOSAR

10

AUTOSAR is the abbreviation of **AUT**omotive **O**pen **S**ystem **AR**chitecture. It is a consortium of automobile manufacturers, suppliers, tool manufacturers, and development service providers who standardize software architecture, interfaces, and the development methodology for the development of software for control units for motor vehicles.

The topic of timing plays a role at various points in AUTOSAR. For example, the AUTOSAR Timing Extensions (TIMEX for short) allow a detailed description of runtime requirements. However, TIMEX is only used by a minority of mass-production projects.

The manner in which the Run-Time Environment (RTE) is configured and used has large influence on the runtime behavior of the software. There are a few pitfalls here that have already caused many a project to stumble.

ARTI, the AUTOSAR Run-Time Interface, is intended to significantly improve tracing and debugging of AUTOSAR modules and the application software of an AUTOSAR ECU. However, ARTI is still in its infancy. The same applies to the topic of timing with the, still quite young, AUTOSAR Adaptive Platform (AP), which is based on POSIX operating systems. Apart from the timing parameters defined in POSIX, important timing parameters for automotive software remain missing, e.g. the Delta Time (DT), which is a central parameter for controllers.

The following chapter will convey the AUTOSAR basics relevant for timing while also offering some practical tips. The aim in both cases is to improve the runtime behavior of AUTOSAR-based ECUs by making them more efficient, more reliable, and safer.

Before 2016 there was basically only *one* AUTOSAR whose standards focused on ECUs implementing real-time operating systems. The AUTOSAR OS standard was a direct successor of the OSEK/VDX standard (see Section 3.2). However, it was already clear that this classic AUTOSAR would be difficult to extend to support high-performance ECUs, such as those required for autonomous driving. Instead of extending the existing group of standards, which would have resulted in a high level

© The Author(s), under exclusive license to Springer Nature Switzerland AG 2021
P. Gliwa, *Embedded Software Timing*,
https://doi.org/10.1007/978-3-030-64144-3_10

Figure 102 The core
'elements' of AUTOSAR

of complexity, a new group of standards was created: AUTOSAR Adaptive Platform
(AP). The previous AUTOSAR became the classic standard, AUTOSAR Classical
Platform (CP). All common elements, such as those concerning the methodology,
were transferred to the newly created Foundation (FO) (see Figure 102). The colors
used for CP, AP, and FO in the figure are maintained throughout this chapter in order
to identify the corresponding group of standards.

The following sections describe the AUTOSAR methodology only using
extracts, namely focusing on the elements relevant to timing. Some of the examples
are taken from a poster [38] that compares the topic of timing in AUTOSAR CP
and in AUTOSAR AP on a single DIN-A1 page. The poster is also available as part
of the online material accompanying this book.

Both AUTOSAR CP and AUTOSAR AP record all configuration information
using ARXML, the AUTOSAR XML format. In ECU projects the information
is distributed over several ARXML files. Even though XML is, in principle, also
readable by humans, this hardly applies to ARXML. The large number of references
and the sheer size of ARXML files—often well over 200 MB for a single ECU—
make it imperative that tools are used to process them.

10.1 AUTOSAR Classical Platform (CP)

How is a clearly defined functionality, such as the idle speed control of an engine
control unit, implemented in a AUTOSAR CP project? This will be clarified in the
following section. To differentiate it from the *functions* programmed in C, the term
functionality is used when dealing with a specific feature.

Figure 103 illustrates two functionalities on their way through the development
process towards execution at runtime. The different levels that are passed through
are highlighted in particular. These levels are not unlike the abstraction levels in
the V-model. The higher a level, the less concrete it is and the more abstract the
relationships are.

10.1.1 Functional Architecture

At the top level is the functional architecture and, at this stage, it is not even defined
whether a functionality is ultimately implemented on an AUTOSAR CP or an
AUTOSAR AP ECU. For this reason, the function architecture level is highlighted

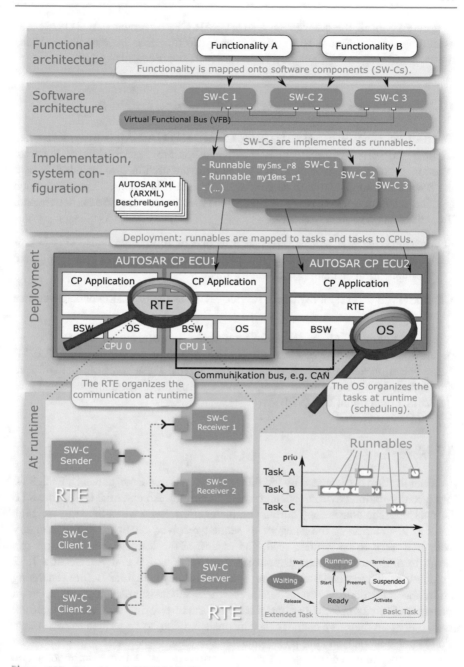

Figure 103 Overview of AUTOSAR CP

in gray and not in the color green (AUTOSAR CP) or brown (AUTOSAR AP) as introduced with Figure 102. Although only two functionalities are shown in the figure, a real system will include a wide variety of functionalities.

10.1.2 Software Architecture, SW-C Definition and VFB

The decision whether to use AUTOSAR CP or AUTOSAR AP is then made when creating the software architecture. Functionalities are described using software components (SW-C, sometimes SWC). The software architecture also defines how the SW-Cs communicate with each other. Although it is not (yet) specified which physical medium is used to exchange data, the type of communication is determined. Two different types are possible:

Sender/Receiver A sender sends data to one or more recipients. This is uni-directional communication. If the receivers need to respond, an additional communication channel is required for this.

Client/Server A client requests a service—such as the provision of data—from a server.

A software component can therefore be a sender or receiver, or client or server, or even take on several of these roles. Or it does not perform any AUTOSAR communication and, therefore, does not have any of these roles. At the software architecture level, the software components communicate via a single virtual bus, the Virtual Functional Bus (VFB).

10.1.3 RTE

Communication is organized by the RTE [39] (**R**un-**T**ime **E**nvironment). Each RTE has a configuration environment and a code generator. An essential task of the RTE is to ensure data consistency during communication. Even when thinking at a high level of abstraction in the category of sender, receiver, client, or server, memory cells will need to be written, read, and copied at the lowest level. That this can result in problems has been demonstrated in Sections 2.9 and 7.3. Section 7.3.2 emphasized that mechanisms ensuring data consistency take time. Before generating code, the RTE code generator analyzes where inconsistent data could occur in principle. For example, if two tasks have the same priority, they cannot interrupt each other and therefore cannot cause inconsistent data to be written to each other. Following this analysis, the RTE generates the data, copies, access functions and copy routines required for secure communication. In places where, in principle, no data consistency problems can occur, efficient, fast mechanisms are used, such as simply global variables.

10.1.4 Implementation, System Configuration, and Runnables

The implementation of a software component is performed by means of *Runnables*. These are usually `void-void` functions, i.e. functions without parameters and without a return value. Runnables have certain requirements, especially with regard to scheduling. One requirement could be "must be executed every 10 milliseconds".

The runnables are assigned to tasks, whereby there are two approaches here. Either the user implements the configuration manually, or parts of the configuration are undertaken by the RTE, or, more precisely, by the RTE configuration environment and the RTE code generator. Without manual intervention, the runnables are usually loaded into a single ECC task and triggered by cyclical events. Section 6.2 provided an impressive display of where blind trust in the default configuration of the AUTOSAR environment can lead.

The tasks themselves are assigned to an *OS Application* and each OS application is assigned to a CPU. The CPU is part of a processor, which in turn is soldered into an ECU. This results in a hierarchy, as already shown in Figure 39.

10.2 AUTOSAR Adaptive Platform (AP)

As already undertaken in the section on the classic AUTOSAR platform, we can now tackle the big-picture view for AUTOSAR AP, which will illustrate the path of a functionality through the development process towards execution during runtime. The overall picture is shown in Figure 104.

AUTOSAR AP is based on POSIX operating systems, whereby the interface between an AUTOSAR AP application and the operating system is limited to the subset PSE51 [40] of the POSIX standard (see also Section 3.4). This means, for example, that a file system is not supported.

10.2.1 Functional Architecture

As with AUTOSAR CP, the definitions of the individual functionalities are independent of the platform used. The fact that only a single, simple functionality has been represented should not obscure the fact that a real system will have a multitude of functionalities.

10.2.2 Software Architecture AA

In AUTOSAR AP, a functionality is distributed to one or more *adaptive applications* (AA). Conversely, it is possible to implement several functionalities with a single AA.

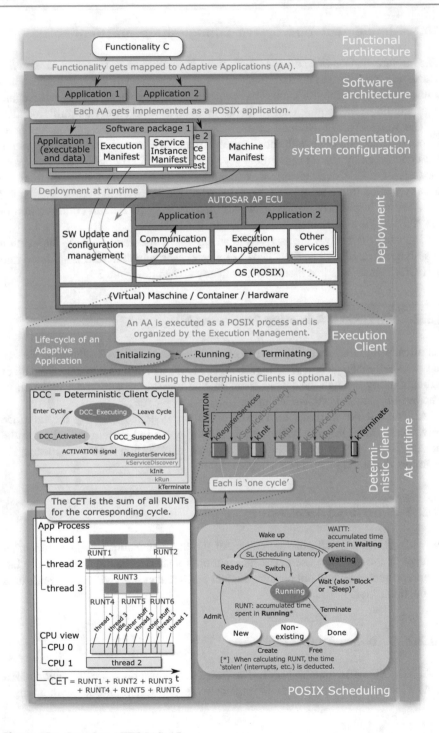

Figure 104 Overview AUTOSAR AP

But what exactly is an adaptive application? For developers who were previously at home using the classic AUTOSAR or in the field of embedded in general it may help to think of an application as it is developed for the PC. Listing 53 shows the main function of an AA that, at first glance, looks very similar to an application for a PC. Unlike classic embedded projects, there is not just one single main function. Instead, each AA has its own main function, in the same way that every C or C++ program developed for the PC has its own main function.

Listing 53 Body of an adaptive application (AA)

```
1  int main(int argc, char *argv[])
2  {
3      int retval = 0;
4      // initialize App data here
5
6      ExecutionClient.ReportExecutionState(kRunning);
7
8      // call App code here (which may or may not return), e.g.:
9      retval = AppCode();
10
11     ExecutionClient.ReportExecutionState(kTerminating);
12     // save persistent App data and free all resources here
13
14     return retval; // terminate
15 }
```

Underlined are the two function calls that turn the main function from Listing 53 into AA—more about this later. The actual application code is not visible as it is hidden behind the function AppCode.

At runtime, the POSIX operating system will treat the AA as a regular *process*. As such it has one or more threads.

Although no code is implemented at the software architecture level, this small code example makes tangible what an adaptive application is at its core.

10.2.3 Implementation and System Configuration

In addition to the actual application, the executable file, *Manifests* must be created at the implementation and system configuration level.

Execution Manifest   This manifest, required for each AA, describes the requirements for the execution of the application and any dependencies on other AAs.
It is not unlike the service description files for systemd on Linux. Here is an
example to illustrate this. Let's assume that, for a home automation system,
software is to be developed for a Raspberry Pi that links radio-controlled power
sockets, radio thermostats, and other devices with the LAN. For this purpose a
daemon is developed (a program that runs permanently). It listens to the radio
traffic and forwards received messages to the LAN for processing by other
devices. Furthermore, it is assumed that the daemon uses a web server.

It is obvious that, when the system boots, first the network drivers, then the web server, and only then the daemon should be started. Amongst other things, it is these kinds of dependencies that are recorded in the `systemd` service description files (`/etc/systemd/system/*.service`).

The execution manifest for AUTOSAR AP also specifies when, and in what form (once or permanently, like said daemon), the AA should be executed.

Service Instance Manifest   This manifest describes the services that the AA uses.

Machine Manifest   All information concerning the executing environment—the concrete hardware, virtual machine, or container—that is independent of an AA is summarized in the Machine Manifest.

As shown in Figure 104, the Execution Manifest and the Service Instance Manifest belong to exactly one AA. Together with the executable and its data (such as parameter sets) they form a *software package*.

The contents and formats of the three manifests are described in detail in the AUTOSAR document "Specification of Manifest" [41].

10.2.4 Deployment

Unlike in the deployment of AUTOSAR CP, the distribution of the software to the ECUs with AUTOSAR AP can be implemented at runtime. It is exactly this ability or property that makes AUTOSAR AP *adaptive*. Over time, the system can adapt to its environment by loading new Adaptive Applications.

The software update and configuration management [42] handles the inclusion of an AA into the system at runtime.

Other important elements are the Communication Management and Execution Management components.

10.2.5 Execution Management and Execution Client

As mentioned above, the instructions underlined in Listing 53 are mandatory for an AA. The call with the parameter `kRunning` marks the beginning of the execution phase of the AA, its 'running' state, and the call with the parameter `kTerminating` marks its end. The two statements inform the Execution Management [43] of the subsequent state of the application.

Whether the running state is left again, or whether this only occurs when the control unit is shut down, depends on the AA.

The following sections look more closely at the running state and what happens to the process more closely. As in the entire chapter, we will focus on those aspects that are related to timing.

10.2.6 Deterministic Execution and Deterministic Client

ECU software based upon the classic AUTOSAR typically has a large proportion of cyclical tasks. Often the software architecture is significantly oriented around the main control algorithm of an application. Periodically activated tasks are, thus, an important part of the developer's everyday work with AUTOSAR CP.

When a AUTOSAR CP developer comes into contact with AUTOSAR AP for the first time, they may wonder how to configure the equivalent of the old periodic tasks in AUTOSAR AP. A POSIX expert (maybe from the Linux environment) who is confronted with this question may shrug his shoulders and answer: Why don't you set a timer yourself? Even if this view is perfectly legitimate, the developer will surely wish for a little more support from the basic software.

The Deterministic Client within AUTOSAR offers the developer a whole bunch of features but, in the following, only those that are relevant for timing will be considered. This includes the capability to execute code periodically—without having to set up a timer yourself.

10.2.6.1 Redundant Execution

It is possible to execute safety-relevant processes a second time in parallel, known as redundant execution. The underlying idea is the same as for the lock-step multi-core (see Section 7.1.2).

10.2.6.2 Periodical Execution

Periodic execution may turn out to be somewhat different than one would expect. First of all, the Deterministic Client requires the AA to comply with the intended state model. Accordingly, an AA can be in one of the following states:

Register Services   – kRegisterServices: The application registers its communication services. That is, it tells the system which communication services it will offer.

Service Discovery   – kServiceDiscovery: The application determines which services will be made available to it.

Init   – kInit: The application initializes itself and its data.

Run   – kRun: The application executes one cycle of its regular code. This is the only state in which code is periodically executed. All other states are to be regarded as 'special cases'. Only the state Service Discovery may occur again in between, if necessary.

Terminate   – kTerminate: The application is preparing for its termination.

A sequence of states may look like this (if no unscheduled service discovery occurs): Register Services → Service Discovery → Init → Run → Run → Run → Run → Run → Run → (...) → Run → Terminate

As mentioned above, it is possible that during the repeated execution of the Run states that a Service Discovery occurs a second time, or even more often. Figure 104 shows such a sequence.

What does this all look like then in code? Listing 54 shows the structure that can be used for an AA using the Deterministic Client.

Listing 54 Implementation of `AppCode()` as a deterministic client

```
1  // Top-level application function of an Adaptive
2  // Application (AA) which uses the Deterministic Client
3  int AppCode(void)
4  {
5      ActivationReturnType dccType;    // Deterministic Client
6      // Cycle (DCC) type
7
8      while (1) { // endless loop
9          dccType = DeterministicClient.WaitForNextActivation();
10         // each execution of the code below is one "Cycle"
11         switch (dccType) {
12         case kRegisterServices:
13             // call handler registering services here
14             break;
15         case kServiceDiscovery:
16             // call service discovery handler here
17             break;
18         case kInit:
19             // call init handler here
20             break;
21         case kRun:
22             // call cyclic App handler here
23             break;
24         case kTerminate:
25             return 0; // terminate with success
26         default: // invalid return value
27             return 1; // terminate with error
28         }
29     }
30 }
```

First of all, it can be seen that the actual code of the application is enclosed by an endless loop. The body of the loop is thus executed again and again, whereby the call of the method `DeterministicClient.WaitForNextActivation()` is always placed at the beginning of each execution. In the state diagram in Figure 104 at the level Deterministic Client, the call of the method corresponds to the transition 'Leave Cycle'.

The method is blocking, meaning that it does not return until an activation by the Execution Manager has taken place. As long as it is blocking, the AA is in a kind of dormant state. After an activation—see transition "ACTIVATION signal"—the POSIX operating system will set the AA a short time later to the POSIX state 'Running' (more on POSIX states shortly). As a result the call `DeterministicClient.WaitForNextActivation()` returns, which corresponds to the transition 'Enter Cycle'. The return value indicates the state of the AA. Depending on this value, the corresponding code can now be executed (see **switch-case** statement in Listing 54). The coloring of the different values in Listing 54 can also be found on the timeline in the Deterministic Client representation in Figures 104 and 105.

The timing parameters shown in the figure for the Deterministic Client are described in the same way. But, first, the levels of the POSIX scheduling need to be reviewed more closely.

10.2.7 POSIX Scheduling

Almost everything shown in Figure 104 at this stage is not AUTOSAR specific but applies generally to POSIX operating systems. The state diagram, as well as the execution of the threads plotted over time, was already presented in Section 3.4.

Once again as a reminder: a POSIX application consists of a process with one or more threads. The threads of a process can run simultaneously on different CPUs of a (multi-core) processor.

Figure 104 shows three threads running on two CPUs. Thread 2 runs exclusively on CPU 1, while thread 1 and thread 3 must share CPU 0 with other applications (the 'other stuff' in the diagram).

10.2.8 Timing in AUTOSAR AP

If one compares the overview diagrams for AUTOSAR CP (Figure 103) and AUTOSAR AP (Figure 104) it is noticeable that, with CP, there is only one state diagram, whereas with AP there are three of them: the Execution Client, the Deterministic Client, and the POSIX Scheduling. The timing parameters at the POSIX level were already covered in Section 4.1.2.

In general, timing parameters are practically always coupled to state machines. Tasks, interrupts, processes, or threads change their states at runtime and timing parameters are usually defined as the time difference between the entry points of two transitions.

So, what does this mean for the timing parameters now that AUTOSAR AP now has three state machines? Which one is used for the definition of the timing parameters? As always, it depends on the point of view. If the view of the Deterministic Client is of interest, its state machines must be used. If the processing of threads is at the center of attention, the POSIX state machine must be used. All state machines are valid simultaneously and produce 'their own' timing parameters.

10.2.8.1 The Timing Parameters of the Deterministic Client

As mentioned above, the use of the Deterministic Client is optional. However, the following definitions of timing parameters can also be applied to non-Deterministic-Client applications and even to non- AUTOSAR projects in general. The definitions are not included in the AUTOSAR standard as of summer 2020. Instead, they can be considered a proposal for a possible set of sensible definitions of timing parameters.

Figure 105 represents a detailed view of the Deterministic Client level from Figure 104. It shows seven cycles of a Deterministic Client. Each cycle corresponds to one execution of the loop body from Listing 54. Applications that do not use the Deterministic Client will still have a loop that will run over and over again—until the application finally terminates.

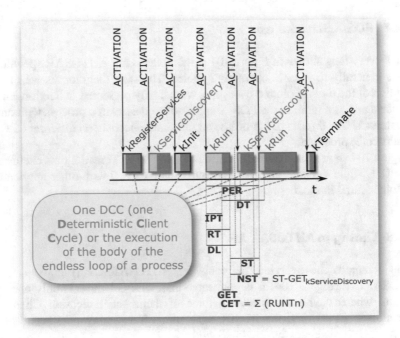

Figure 105 The timing parameters of the Deterministic Client

For each cycle type, kRegisterServices, kServiceDiscovery, kInit, kRun, and kTerminate, the timing parameters are considered separately. In Figure 105, all the timing parameters shown refer to the kRun cycle. In what follows, the timing parameters will be defined and described in detail. The definitions are deliberately very close to the definitions of the timing parameters in the RTOS environment as presented in Section 4.1.1. In fact, portability was an essential aspect of the following definitions.

The idea behind it is obvious, namely to ensure that *one* definition of timing parameters exists that is applicable to both worlds—to the RTOS world and AUTOSAR CP on the one side, and to the POSIX world and AUTOSAR AP on the other.

PER—PERiod

The period describes the time difference between two successive activations of the same cycle type. It should correspond to the configured period at runtime.

DT—Delta Time

The delta time describes the time between the start of an instance and the start of a subsequent instance of the same cycle type.

JIT—Jitter

The jitter describes the deviation of an actual cycle time from the desired cycle time. It is not shown in Figure 105 but it was discussed in detail in Section 4.1.1 and everything said there is also applicable to the jitter of the Deterministic Client.

J—Absolute Jitter

Like the jitter, the absolute jitter was discussed in detail in Section 4.1.1 and everything said there also applies to the absolute jitter of the Deterministic Client.

IPT—Initial Pending Time

The initial pending time is the time a cycle 'waits for its start', that is, the time difference between activation and start or, more precisely, the time difference between the ACTIVATION event and the return from the call to `DeterministicClient.WaitForNextActivation()`.

RT—Response Time

The response time indicates how much time elapses between the ACTIVA-TION event and the end of the affected loop body, i.e. until the call to `DeterministicClient.WaitForNextActivation()`.

DL—DeadLine

The deadline is nothing more than the maximum allowed response time. A deadline is therefore a specification; deadlines cannot be measured.

ST—Slack Time, remaining time

The slack time describes the 'gap' between the end of one loop pass and the ACTIVATION event of the next loop body of the same cycle type. It is irrelevant whether a loop pass of another cycle type takes place in this gap. According to the AUTOSAR specification, only one `kServiceDiscovery` can do this between two `kRun` Instances.

If a `kServiceDiscovery` instance inserts itself between two `kRun` instances at runtime, the slack time can only be used to a limited extent for determining the amount of time additional code may consume before causing a timing problem. For such an analysis it is better to use the NST described below.

NST—Net Slack Time, net remaining time

The net slack time is calculated from the slack time minus all GETs that fall within the slack time period, and that belong to a loop run of another cycle type. As said, this is only possible for `kServiceDiscovery`. Figure 105 illustrates such a situation.

GET—Gross Execution Time, gross runtime

The gross execution time is calculated from the time difference between the start time and end time of a loop pass, i.e. from the return of the function `DeterministicClient.WaitForNextActivation()` until it is called again.

CET—Core Execution Time, net runtime

As in the RTOS environment, AUTOSAR CP and POSIX applications should generally reflect the CET that the processor spent processing the code under consideration. Since a process can start several threads, and since these can be executed simultaneously on different CPUs of the same processor, the runtimes of all threads must be added together. Basis for this calculation are the RUNT times provided by POSIX. Figure 104 illustrates this for threads 1–3.

It should be noted that the result of this summation, when several threads are processed in parallel, can deliver a CET for a loop pass that is larger than the GET for the same loop body.

10.3   TIMEX (AUTOSAR Timing Extensions)

The AUTOSAR Timing Extensions, TIMEX for short, were introduced with AUTOSAR Release 4.0. The goal was, and still is, to formally describe timing aspects and, above all, to be able to specify formal timing requirements.

TIMEX uses AUTOSAR XML as a description format, which means that even quite simple timing requirements such as "The response time of TASK A must not be greater than two milliseconds" result in many lines of ARXML code. Such code can only be interpreted, or even created, by experienced TIMEX experts. This would not be a big problem if there were tools that would relieve the developer of the cumbersome handling of ARXML. Unfortunately, in more than 10 years that TIMEX has been available as a standard, none of the tool providers has tackled the development of such a tool.

Section 9.1.1.2 highlighted an alternative or interim solution if TIMEX is not practical for a AUTOSAR project for the reasons mentioned above.

Nevertheless, an introduction to the TIMEX specification will now follow.

10.3.1 Goals

TIMEX pursues two goals with the definition of timing requirements. On the one hand, it supports the configuration of a system so that configuration decisions best serve the timing requirements. On the other hand, TIMEX enables the verification of whether the timing requirements have been met. The idea behind this second goal has already been discussed in various sections of this book, so we will now go into the first goal in more detail.

Assuming that timing requirements were specified fully and formally for all software components of a project, automatic configuration of operating systems,

and automatic distribution of runnables to tasks—as well as tasks to CPUs—
would be conceivable. Although today's AUTOSAR tools are still far from this,
the availability of all timing requirements in a precise form is the first step towards
a completely automated configuration.

10.3.2 Events and Event Chains

So, what can timing requirements specified with TIMEX apply to? Basically, they
can be applied to *events* and *event chains*. Events are all AUTOSAR events that are
uniquely identifiable, such as the sending or receiving of data, the requesting of a
service, the receipt of any subsequent feedback in this regard, or the activation, start,
or termination of an executable unit.

Event chains are chains of two or more events. It is often useful to organize these
event chains hierarchically. An example will illustrate this.

Assuming that there is a requirement that the brake lights of a vehicle illuminate
no later than 200 ms after the brake pedal is pressed, this requirement can be mapped
at the top level to an event chain with two events. These are:

- Brake pedal changes from 'not depressed' to 'depressed'.
- Brake lights change from 'off' to 'illuminated'.

The brake pedal will most likely not be connected to the same control unit as the
brake lights. Therefore, at least two ECUs will be involved, possibly even more
if, for example, there is a gateway between these two ECUs. The information that
the brake pedal is depressed (Event 1) will reach the control unit A first. There,
the information received is checked for plausibility (Event 2) and processed further
(Event 3). Event 4 marks the request to send the information, and Event 5 the actual
moment it is sent. In this sense, the process continues until, finally, the brake lights
are illuminated with Event x. The end-to-end requirement that the brake lights
illuminate at the latest 200 ms after the brake pedal is pressed must, therefore, be
broken down into the sections of the active chain.

10.3.3 TIMEX Requirement Types

TIMEX defines different types of requirements (constraints). For each type the
following list provides an example of its application.

EventTriggeringConstraint   Typical use case: monitoring the jitter of a periodic
event.

LatencyTimingConstraint   Typical use case: avoidance of doubly-received or
lost data due to unsynchronized or badly synchronized senders/receivers (see
also Section 6.4).

AgeConstraint   Typical use case: ensuring that data is not too old.

SynchronizationTimingConstraint   Typical use case: synchronization of sub-systems to avoid race conditions.

OffsetTimingConstraint   Typical use case: monitoring a desired time offset between two events.

ExecutionOrderConstraint   Typical use case: monitoring of the desired sequence during the processing of runnables.

ExecutionTimeConstraint   Typical use case: monitoring of the maximum allowed CET, such as that of a runnable.

10.3.4 AUTOSAR/TIMEX Views

AUTOSAR looks at a system from different angles known as *Views*. TIMEX adopts this approach so that a timing request can always be assigned to a specific TIMEX View. In detail these are:

VfbTiming   Timing during the interaction of AUTOSAR software components (SW-Cs) via the Virtual Functional Bus (VFB, see also Section 10.1.2).

SwcTiming   Software component's internal timing.

SystemTiming   Timing across control units.

BswModuleTiming   Basic software module (BSW) internal timing (analogous to SwcTiming).

BswCompositionTiming   Timing during interaction of several basic software modules.

EcuTiming   Timing for a completely configured control unit. The interaction of all software components and basic software modules is clearly defined in this completed configuration. In contrast, the other Views specify timing requirements *independently* of their use in a specific ECU.

10.4   ARTI (AUTOSAR/ASAM Run-Time Interface)

ARTI was launched in 2016 with the aim of significantly simplifying the timing analysis of AUTOSAR projects. The similarity in name to ORTI, the OSEK Run-Time Interface, is intentional. In some ways, ARTI is to be understood as a successor to ORTI, but in others it goes significantly beyond it.

In the course of the development of ARTI it turned out that it does not make sense to organize all ARTI features under the roof of AUTOSAR, since large parts of ARTI are not AUTOSAR specific and can be used for non- AUTOSAR applications. For this reason, an ASAM project was started in early 2019 and completed in February 2020. ASAM stands for "Association for Standardization of Automation and Measuring Systems" and is a registered association that supports and organizes standardization in the field of software development, simulation, and diagnosis, as well as automation and measuring procedures.

Before going into detail on AUTOSAR ARTI and ASAM ARTI, the following list of goals and features should provide an overview of what ARTI is all about. Not all of the points listed therein are covered by the AUTOSAR release of November 2019 and the ASAM release of February 2020.

- Goals and features of AUTOSAR ARTI   **AUTOSAR**
  - Debugging with OS awareness, i.e. with a debugger that is 'aware' of the operating system and thus can, for example, display the states of all tasks. A later version of ARTI will also provide the debugger with information about other AUTOSAR modules so that RTE awareness, COM Stack awareness, and so on, will also be possible.
  - Analog to debugging, tracing can also collect, visualize, and evaluate information specific to the operating system, the COM stack, the RTE, etc.
  - Tracing of runnables.
  - Tracing of events defined by the user. The 'awareness' can thus also be extended to the application.
  - Support of hardware-based tracing.
  - Support of software-based tracing.
  - Profiling based on measurements, either via software instrumentation or via special hardware, such as the performance counters of the Infineon AURIX.
  - Multi-core tracing including the synchronization of the traces of different CPUs.
  - Multi ECU tracing including the synchronization of the traces of different ECUs.
  - Tracing and measurement of TIMEX constraints, i.c. tracing and measurement of the corresponding events and event chains.
  - Support for AUTOSAR AP.
- Aims and features of ASAM ARTI   **ASAM**
  - Standardized trace format for exchanging trace data.
  - Standardized format for exchanging timing parameters.

Figure 106 shows the data flows and tools that play a role in connection with ARTI. In the background of the graphic you can see a V that indicates that the individual steps are oriented toward the V-Model. However, while the assignment in the vertical should not be taken too literally, the horizontal axis provides a clear division into two parts. All steps that take place *before* the availability of the executable file are to be assigned to AUTOSAR, while all steps after that, i.e. the whole subject of verification on the right, belong to ASAM ARTI.

A similar approach has been used in the past for MCD-2 MC. MCD-2 MC is probably known to most developers simply as ASAP2, an ASAM standard for measurement and calibration. Again, AUTOSAR standards refer to an ASAM standard that describes interfaces for tools on the right-hand side of the V-model.

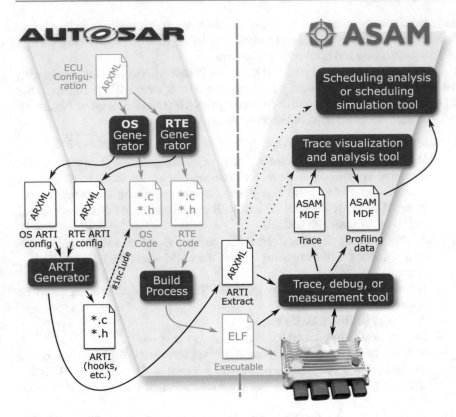

Figure 106 Overview for ARTI

All files shown in Figure 106 in gray are files that existed prior to ARTI in the
AUTOSAR process and have now been extended by ARTI.

10.4.1 AUTOSAR ARTI

The mission of AUTOSAR ARTI consists of creating the prerequisites on the left-
hand side of the V-Model so that runtime measurements can later be made and traces
recorded on the right-hand side.

How does this look in detail? It starts by extending the ECU configuration
using ARTI. The ECU configuration contains all information needed by the code
generators of the individual AUTOSAR modules to generate code. A central aspect
at this point is which AUTOSAR modules will support ARTI in the software later
on.

The ECU configuration is now read by the code generators. For better clarity,
only two of these generators are shown in Figure 106: one for the operating system
and another for the RTE. However, the concept is transferable to all AUTOSAR

modules for which a timing analysis of their services and data flows may be of interest.

10.4.2 ASAM ARTI

AUTOSAR ARTI is used to enable tracing or runtime measurements for a specific software revision in order to collect all necessary information. The information about the system configuration—that is, which tasks exist, which runnables exist, how the runnables are assigned to the tasks, and so on—is written to a new AUTOSAR XML file. In Figure 106 this can be seen as "ARTI Extract".

A trace tool can read this information and read the trace data from the ECU (see Figure 106). Usually it will display the traces themselves, but it can also save them in the ASAM ARTI trace format so that other tools can read in, interpret, and evaluate the trace data.

One result of the evaluation of trace data could be the determination of the minimum, maximum, and average values of all timing parameters (Core Execution Time CET, Response Time RT, etc.). Timing parameters for AUTOSAR CP were addressed in Section 4.1.1, and those for AUTOSAR AP in the Sections 10.2.8.1 and 4.1.2.

The *profiling data* thus obtained can be stored in the ASAM ARTI profiling data format and then serve as input data for a scheduling simulation or a static scheduling analysis.

Up to now and in connection with ARTI we have mostly spoken of tracing. However, pure runtime measurement technology is also supported. In this case, the code called via the ARTI hooks does not generate any trace data but it does calculate the desired timing parameters directly. The profiling data determined at runtime must then be read out in a suitable form, which is usually implemented using PC software designed for this purpose. This PC software can then save the data in the ASAM ARTI profiling data format.

The ASAM specification " ARTI ASAM Run-Time Interface" [44] defines both the format for the trace and profiling data. In both cases, the format is based on another ASAM standard, the Measurement Data Format in its fourth version (MDF4 [45] for short). MDF4 is a binary data format that is designed to efficiently store even large amounts of data for post-processing or permanent storage.

Both the ASAM ARTI trace data, as well as profiling data, can optionally contain information about the system configuration such as the information contained in the AUTOSAR XML file ARTI Extract. For this reason, the arrows from this file to the 'Trace Visualization and Analysis Tool' and the 'Scheduling Analysis or Simulation' are dotted in Figure 106.

10.5   Technical Report "Timing Analysis"

AUTOSAR defines different types of documents. Requirement specifications define the requirements that are placed on standardization documents (specifications). There are documents that are purely explanatory (Explanatory Documents) and also "Technical Reports".

The latter category includes the document with the somewhat unwieldy name "Recommended Methods and Practices for Timing Analysis and Design within the AUTOSAR Development Process" [9], which is often simplified to the "Timing Analysis" document.

After a general introduction it approaches the topic of timing with use cases. These use cases are assigned to different levels within the V-model or different areas of use. In detail these are:

- Design level
- Functional level
- End-to-end timing for distributed functions
- Network level
- ECU level

Furthermore, the document deals with the definition of timing parameters and methods related to timing. This includes the various analysis techniques as discussed in Chapter 5.

Like all AUTOSAR working groups, the creators of the "Timing Analysis" document welcome those interested in participating. The only requirement is an AUTOSAR membership as Premium Partner, Development Partner, or Attendee. Beyond that, all that is required is a keen interest for the topic of timing.

10.6   Summary

In Chapter 10 an overview was provided of the standards and working groups within AUTOSAR that are related to timing.

Attempts were made to apply the timing parameters associated with classic AUTOSAR and OSEK/VDX to AUTOSAR AP. Although the description is based on the Deterministic Client, the presented interpretation of the timing parameters is basically independent of the Deterministic Client and can generally be applied in the POSIX environment.

It remains to be seen whether TIMEX will prove itself in practical, everyday use and whether it will find its way into a large number of projects. Until then, minimum or maximum value specifications of selected timing parameters will allow the simple and pragmatic specification of at least part of the timing requirements. Section 9.1.1.2 in the Chapter 9 had already taken up this idea.

AUTOSAR ARTI is still quite young in comparison to TIMEX and therefore still needs to prove itself in practice. If it is successful, users can look forward to a much improved interaction of AUTOSAR modules—especially the operating system—and timing analysis tools. ASAM ARTI will also considerably simplify the data exchange between the timing analysis tools.

Safety and ISO 26262

11

This chapter deals with safety in the sense of limiting the potential catastrophes that embedded systems could cause. Controlling the cooling system in a nuclear power plant, a pacemaker, the flight controller of an aircraft, or the brake control unit in a vehicle; software errors in such embedded systems can have potentially fatal consequences.

Timing is only one aspect of many in embedded software. However, there can be no safe embedded software without stable and secure timing.

In addition to technical topics, the following chapter also deals with legal aspects. Managers and project managers are confronted with many questions regarding the kind of timing analysis that has to be performed, and what timing protection needs to be implemented in order to be legally well positioned.

Some of the aspects dealt with below are summarized on a DIN-A1 poster "ISO 26262". This poster is available for download as a PDF [46]. It is also available as part of the online material accompanying this book.

11.1 Basics

Safety alone is not everything. Building a very safe system is not that difficult if enough monitors are integrated that can switch the critical functionality of the system into a secure state should one of them be activated. However, one would run the risk of having acquired safety for the price of a very low availability. In extreme cases, the product would be safe but it would no longer actually perform its function.

It is therefore clear that both the error rate and the availability must be defined with appropriate values.

© The Author(s), under exclusive license to Springer Nature Switzerland AG 2021
P. Gliwa, *Embedded Software Timing*,
https://doi.org/10.1007/978-3-030-64144-3_11

11.1.1 Risk

In connection with safety there is always the question of *risk* that emanates from a danger. An attempt can be made to quantify the risk so that it can be classified. In doing so, one encounters various definitions that include two or three aspects that contribute to risk. EN ISO 12100:2010 defines risk as a combination of probability of occurrence and extent of damage [47]. This is usually undertaken by simply calculating the product of:

$$Risk \ = \ Probability \ of \ occurrence \ \cdot \ Extent \ of \ damage \qquad (19)$$

An alternative, which is used in a similar form in ISO 26262, additionally includes the aspect of how well the hazard can be controlled upon entry. Put into a formula, it could look like this:

$$Risk \ = \ \frac{Probability \ of \ occurrence \ \cdot \ Extent \ of \ damage}{Controllability} \qquad (20)$$

11.1.2 SIL: Safety Integrity Level

Virtually every safety standard defines *Safety Integrity Levels*. Each component of an embedded system—covering both hardware and software—is assigned a specific safety integrity level after a corresponding risk analysis. The question being answered here is: How many deaths and injuries a malfunction of the component under consideration could cause if no other safety mechanisms were to take effect?

An example will illustrate this. The electrical components of a brake valve can cause one of the wheels of a vehicle to suddenly brake sharply if they were to fail. If this happens at high speed on the motorway, an accident with several fatalities is easily conceivable.

The control electronics for the interior lighting of the trunk are different. Even with a lot of imagination, you can't envisage a dangerous situation here—a fire due to overheating, for example, is out of the question.

Which Safety Integrity Levels are available depends on the standard used (see Table 8).

11.1.3 Out of Context, in Context, and Proven-in-use

In the vast majority of cases, the software of an embedded system is not 100% created by a single company. This means that, usually, software from several suppliers are integrated into the final product as components and this covers everything from operating systems, drivers, and bootloaders to protocol stacks, individual algorithms, and so on. In the case of these third-party components, the question arises as to how the protection should be carried out. There are basically three possibilities here.

In context   The supplied component is certified in the same way as the self-developed code. While this sounds simple, it is sometimes difficult or even impossible to do. In order to be able to analyze, test, and thus secure supplied code as you would your own, the availability of the source code is almost mandatory. It will therefore be difficult to provide in-context certification for a component for which the supplier does not provide the source code. The certification of a component together with a specific product (in context) is shown in Figure 107.

Out of context   The component is certified independently of a specific project or application. The certification therefore takes place outside of a project context, hence the term 'out of context'.

With ISO 26262 this becomes even more concrete: a *(System) Element out of Context* (SEooC) is a component developed and then certified according to ISO 26262 independent of a specific project.

In addition, there are components that were not originally developed according to ISO 26262, but were later adapted according to the standard and then certified. Such a component is then termed *Commercial Off-The-Shelf* (COTS) software component.

So, does this mean that with SEooC or COTS components the user of the certified component does not need to do anything else? Not quite, because they also have to make sure that the *Safety Manual* created in the course of both their product's and the component's certification is taken into account in their project. The *Safety Manual* can be thought of as a kind of user manual for the safe use of the component. The user must therefore prove that they have established all the necessary boundary conditions for safe operation of the component and that the restrictions imposed by the component have been taken into account (see also Figure 108).

It should also be noted that the effort required for this is considerably lower than for 'in context' protection.

Proven in operation   Finally, the argument can be introduced during validation that a component has already been used many times and over a very long period of time without any problems. In this case, proof must be provided that the boundary conditions for the component do not differ for the current project compared to the projects that have proven themselves in operation.

Sometimes, an older version of the compiler is deliberately used in security-relevant projects because it is tried and tested. In such cases, all compiler errors that have come to light since the release of the old compiler must be examined individually for their relevance to the current project.

Figure 107 In context
certification

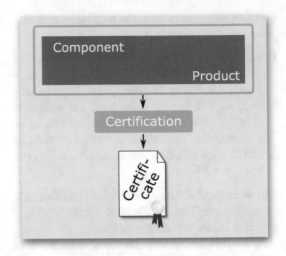

Figure 108 Out of context
certification

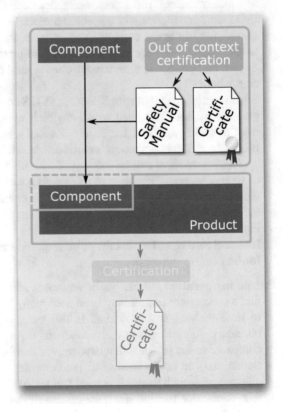

11.2   Safety Standards, Timing, and Timing Verification

Whether aerospace, the medical sector, railways, automotive, or any other field, everyone has their own standards that are relevant to software development for embedded systems. Table 8 provides only a small selection of the safety standards for selected areas. The table also shows which safety integrity levels are covered by each standard and how they are termed.

Table 8 Examples of important safety standards for embedded software

Standard	SIL (from–to)		Description
	Low	High	
IEC 61508	SIL1	SIL4	General safety standard for electrical/electronical components
IEC 61508 Edition 2	SIL1	SIL4	Successor to IEC 61508
ISO 26262*	ASIL A	ASIL D	Safety standard for automotive software
IEC 62061*	SIL1	SIL3	Safety standard for electrical control systems of machines
IEC 62304*	Class A	Class C	Safety standard for the software of medical equipment
DO 178-B	Level E	Level A	Safety standard for avionics software
DO 178-C	Level E	Level A	Succesor to DO 178-B
CENELEC prEN 50128	SIL1	SIL4	Safety standard for railway software

[*] Derived from standard IEC 61508

How do the standards take the issue of timing into account? Overall, very superficially. They require the individual software components to be free of interference. This also includes interference in the time domain.

If a system has to execute a safety-relevant function every millisecond, it will be impeded if another function blocks the interrupts due to a software error and then hangs in an endless loop. This is a simple and striking example; in reality, the impact will usually be less drastic, and therefore harder to find. A function may take significantly longer than usual in rare cases. Perhaps, in equally rare cases, the scheduling will be affected in such a negative way that it will cause a problem with the synchronization of data, which in rare cases can lead to a critical failure of the whole system. This second example outlines a concatenation of three rare problems at three different levels: the code level, the scheduling level, and the communication level.

So, with which timing analysis technique should safety-relevant software be verified? The standards stipulate that tests are unavoidable. Depending on the safety integrity level, static analysis methods may also be recommended. As a rule, recommendations are made in the form of tables. For the various safety integrity levels it will be indicated for each specific technique whether it is strongly recommended, recommended, or not required.

The analysis techniques that are ultimately used must be decided individually for each project. An early discussion with experts from certification authorities is very helpful here and ensures that the desired level of assurance is both sufficient but not disproportionate, and thereby not unnecessarily expensive.

Back to the example with the concatenation of three rare problems on three different levels. With our knowledge of the timing analysis techniques presented in Chapter 5, the WCET of the function could be determined by static code analysis. The result, along with the WCETs of many other functions, could then be used as input for a static scheduling analysis. Using scheduling simulation, the synchronization during data exchange could be simulated and, finally, compared to the previous, more theoretical steps, tracing would ensure that the real system actually behaves as expected.

Very few projects will use this full range of timing analysis techniques. A good knowledge of the requirements of the standard, the system to be protected, and the timing analysis techniques is a prerequisite for finding the optimal balance in the end.

Experience has shown that another aspect is at least as important for safe and reliable timing: simplicity of architecture and scheduling. As simple and short as the statement "Keep it simple!" is, it is also fundamentally important and the correct thing to do.

11.3   Tools for Timing Verification

Tools being used to verify the timing of software, are also software. Like practically all software, the tools used for timing assurance are not free of errors. This means that there is a risk that a tool will give the green light for an aspect of timing verification even though there is a timing problem (false positive).

The safety standards do not prescribe the use of specific products but do require an assessment of the *reliability* of each tool used for verification. As with the risk analysis presented above, the evaluation of tools requires an estimation of the probability of failure and an analysis of the possible impact of that failure. The result is the tool confidence level (TCL).

As with risk, controllability can also be taken into account here, such as by using a second method for verification. It is very unlikely that an error will be found in both methods at the same time in the same analysis.

As a result of such a tool analysis a *qualification* of the tool may be necessary. A formal qualification is comparable to a certification. An embedded system is *certified* according to the corresponding standard (see Table 8). Tools that are involved in the generation of software, such as compilers and code generators, as well as tools used for verification, are *qualified*.

11.4 Legal Aspects

Product liability states that the producer of a product can be held liable for any danger posed by the product. If significant damage is caused due to a fault of the product—such as an accident with several deaths—the manufacturer of the product will have to answer in court. If it turns out that the causal defect was within a supplied component and the manufacturer of the product used the component completely correctly, the responsibility is transferred to the supplier of the component. The following considerations therefore concern either the manufacturer of the product or the manufacturer of the component.

It is also obvious that the care taken by manufacturers in the development of the product or component has a significant influence in determining the penalty. Everyone has heard of the terms 'negligent' or 'grossly negligent'. In such court cases it is not possible to determine precisely in advance according to which criteria a classification will be made, and much depends on the country, the court, the respective situation, and other factors.

What has crystallized over time is that development should be undertaken according to the *state of the art*.

The *state of the art* defines a technical level that can be considered the recommended standard. The book "Engineering Management" describes the relevance of the state of the art in the context of liability: "An engineer may defend against a claim of negligence by contending that they met the standards of their profession and the state of the art" [48].

What does this mean in concrete terms for verifying the timing of embedded systems? As soon as an analysis technique or method has been used several times in the development of a product, and if it has been reported publicly—for example in lectures, at congresses, or in technical journals—it can be assumed that this analysis technique or method is state of the art.

By the way, German speaking countries differentiate between "Stand der Technik" (state of the art) and "Stand der Wissenschaft" which could be translated as "state of research". With respect to liability, the "Stand der Technik" is that which should be pursued during development [49].

11.5 Summary

In this last, rather short, Chapter 11, having clarified the terms 'risk' and 'safety integrity level' and explained the different types of certification, the timing-relevant aspects of existing safety standards were examined.

With regard to product liability, every project manager is advised to operate timing safeguards in their project to at least the level of 'state of the art'.

Outlook

12

What role will timing play in software development over the next few years? And what will be the role of timing analysis?

The moment in time at which software writes itself and makes the craft of software development superfluous is still a while away. At the same time, the trend of embedded systems deepening their penetration into our world remains unwavering. Not only is the number of embedded systems constantly increasing, the software within those systems is becoming ever more important. One can observe this very clearly within the automotive industry. For many years, engine management ECUs were the only components with software in them but, in recent years, the number of control units has only continued to increase. We have reached the point today where luxury-class vehicles have well over one hundred microprocessors upon which embedded software is executed. With the introduction of autonomous driving, we can expect that the amount of software required to implement a vehicle will see a significant jump.

The correct execution of the software and especially the correct execution with respect to *timing*—i.e. solid embedded software timing—is playing an increasingly important role. Building knowledge, skills, tools, and infrastructure in this environment is therefore an investment in the future.

So, which technical innovations are to be seen on the horizon?

The expected improvement in data exchange between timing analysis tools via ASAM ARTI in itself is not yet a technical innovation. However, it will ensure that more timing analysis is undertaken.

It is clear that the interfaces to hardware-based tracing are not improving or expanding to the same extent as the number of CPUs being added to processors, not to mention the increase in clock speeds being offered. Thus, the trend towards more timing analysis based on software instrumentation will continue. Only then will it be possible to use the same technology everywhere: in the laboratory, at the test site, in the real product environment—for example in a vehicle—as well as for timing monitoring as part of the safety concept in the final product.

© The Author(s), under exclusive license to Springer Nature Switzerland AG 2021   297
P. Gliwa, *Embedded Software Timing*,
https://doi.org/10.1007/978-3-030-64144-3_12

Static scheduling analysis and scheduling simulation will certainly play an increasingly important role. For systems whose configuration may change after delivery—'over-the-air updates' is a hot topic—it is conceivable that the embedded system itself performs some simple timing analyses. Even *before* installing new software components, such system checks could determine whether they will endanger the timing of the system. If this were found to be the case, the update would be rejected. If necessary, space would have to be created in the scheduling by removing other software components that are no longer required.

There is great potential in an extensive use of parallelism at the functional level. Today's code generators lack the required knowledge of the system configuration, i.e. information on how many CPUs are available, how their scheduling functions, how they communicate with each other, and how they synchronize with each other. A development environment that combines the functional model, system configuration, and code generation under one roof could produce very efficient, reliable, and safe software and, at the same time, drastically reduce the development effort.

Embedded software timing will become even more important in the coming years. Blessed are those who can demonstrate solid knowledge and corresponding skills in this environment.

References

[1] Wikipedia *Article "V-Model", 2020* https://en.wikipedia.org/wiki/V-Model
[2] Steve Chamberlain, Ian Lance Taylor *The GNU linker, Section 3.1 "Basic Linker Script Concepts", 2010* https://www.eecs.umich.edu/courses/eecs373/readings/Linker.pdf
[3] Microchip Technology Inc. *AVR® Instruction Set Manual, 2016* https://www.microchip.com
[4] Wikipedia *Article "Binary prefix", 2020* https://en.wikipedia.org/wiki/Binary_prefix
[5] OSEK/VDX *Operating System Specification 2.2.3, February 2005*
[6] GLIWA GmbH embedded systems *gliwOS—Embedded Operating System, User Guide Version 1.4, January 2019*
[7] Wikipedia *Article "Ada (programming language)", 2020* https://en.wikipedia.org/wiki/Ada_(programming_language)
[8] IEEE *Standardized Application Environment Profile (AEP)— POSIX Realtime and Embedded Application Support, 2004* https://icccxplore.iccc.org/document/1342418
[9] AUTOSAR *Technical Report 645 "Timing Analysis": Recommended Methods and Practices for Timing Analysis and Design within the AUTOSAR Development Process, November 2019* https://www.autosar.org/standards/classic-platform
[10] GLIWA GmbH embedded systems *Timing Analysis Techniques, 2013* https://gliwa.com/downloads/Gliwa_TimingPoster_Web.pdf
[11] Schloss Dagstuhl, Leibniz-Zentrum für Informatik GmbH *The Logical Execution Time Paradigm: New Perspectives for Multicore Systems, 2018* https://drops.dagstuhl.de/opus/volltexte/2018/9293/pdf/dagrep_v008_i002_p122_18092.pdf
[12] AbsInt Angewandte Informatik GmbH *aiT—The industry standard for static timing analysis, 2020* https://www.absint.com/ait
[13] AbsInt Angewandte Informatik GmbH *Website, 2020* https://www.absint.com
[14] AbsInt GmbH *Obtaining Worst-Case Execution Time Bounds on Modern Microprocessors, 2018* https://www.absint.com/papers/2018_ew_tw.pdf
[15] Synopsys Corporate *Virtual Prototyping, Website, 2020* https://www.synopsys.com/verification/virtual-prototyping.html
[16] Synopsys Corporate *Website, 2020* https://www.synopsys.com
[17] Microchip Technology Inc. *Atmel® Studio 7, 2020* https://www.microchip.com/mplab/avr-support/atmel-studio-7
[18] Altium Limited *TASKING Embedded Profiler for TriCore/ AURIX, 2020* https://www.tasking.com/products/tasking-embedded-profiler
[19] OSEK/VDX *OSEK Run Time Interface (ORTI) Part A: Language Specification, November 2005*
[20] iSYSTEM AG für Informatiksysteme *winIDEA—IDE, Debug and Trace Tool, 2020* https://www.isystem.com/products/software/winidea.html
[21] Lauterbach GmbH *Website, 2020* https://www.lauterbach.com
[22] iSYSTEM AG für Informatiksysteme *Website, 2020* https://www.isystem.com

[23] GLIWA GmbH embedded systems *Analysis suite T1, 2020* https://gliwa.com/index.php?page=products_T1

[24] AUTOSAR *Specification of Synchronized Time-Base Manager, November 2019* https://www.autosar.org/standards/classic-platform

[25] AMALTHEA *Project Website, 2020* http://www.amalthea-project.org

[26] Vector Informatik GmbH *TA Tool Suite Website, 2019* https://www.vector.com/int/en/products/products-a-z/software/ta-tool-suite/

[27] INCHRON GmbH *chronVAL Website, 2020* https://www.inchron.com/tool-suite/chronval

[28] Wikipedia *Article "Butterfly effect", 2020* https://en.wikipedia.org/wiki/Butterfly_effect

[29] INCHRON GmbH *Website, 2020* https://www.inchron.com

[30] AUTOSAR *Specification of Timing Extensions, November 2019* https://www.autosar.org/standards/classic-platform

[31] Oliver Scheickl und Christoph Ainhauser, BMW Car IT GmbH, Peter Gliwa, GLIWA GmbH *Tool Support for Seamless System Development based on AUTOSAR Timing Extensions, 2012* https://gliwa.com/downloads/ERTS2012_ARTimingTooling_final.pdf

[32] GLIWA GmbH embedded systems *An Introduction to Automotive Multi-core Embedded Software Timing, 2015* https://gliwa.com/downloads/Gliwa_MulticorePoster_Web.pdf

[33] Texas Instruments Inc. *Texas Instruments TMS570 architecture, 2015* https://www.ti.com

[34] IAR Systems AB *IAR Application Note G-002, Fast square root in C, 2020* http://netstorage.iar.com/SuppDB/Public/SUPPORT/000419/AN-G-002.pdf

[35] Infineon Technologies AG *AURIX TC27x D-Step, December 2014* https://www.infineon.com

[36] Wikipedia *Article "Duff's Device", 2020* https://en.wikipedia.org/wiki/Duff%27s_device

[37] Wikipedia *Article "Chris Rupp", 2020* https://de.wikipedia.org/wiki/Chris_Rupp

[38] Peter Gliwa *Poster "Timing: AUTOSAR CP, AUTOSAR AP, POSIX", February 2019* https://gliwa.com/downloads/Gliwa-POSIX-Poster_Web.pdf

[39] AUTOSAR *Specification of RTE Software, November 2019* https://www.autosar.org/standards/classic-platform

[40] IEEE Standards Association *IEEE 1003.13-2003 "Standardized Application Environment Profile - Realtime and Embedded Application Support", 2003* https://standards.ieee.org/standard/1003_13-2003.html

[41] AUTOSAR *Specification of Manifest, November 2019* https://www.autosar.org/standards/adaptive-platform

[42] AUTOSAR *Specification of Update and Configuration Management, November 2019* https://www.autosar.org/standards/adaptive-platform

[43] AUTOSAR *Specification of Execution Management, November 2019* https://www.autosar.org/standards/adaptive-platform

[44] ASAM *ASAM Run-Time Interface, February 2020* https://www.asam.net/standards/detail/arti

[45] ASAM *Measurement Data Format Version 4.2.0, September 2019* https://www.asam.net/standards/detail/mdf

[46] GLIWA GmbH embedded systems *ISO26262, 2017* https://gliwa.com/downloads/Gliwa_ISO26262_Poster.pdf

[47] Wikipedia *Article "Risk", 2020* https://en.wikipedia.org/wiki/Risk

[48] David I. Cleland, Dundar F. Kocaoglu, Julienne Brown *Engineering Management, page 440, 1981*

[49] Wikipedia *Article "Stand der Technik", Section "Deutschland", 2020* https://de.wikipedia.org/wiki/Stand_der_Technik#Deutschland

Index

© The Author(s), under exclusive license to Springer Nature Switzerland AG 2021
P. Gliwa, *Embedded Software Timing*,
https://doi.org/10.1007/978-3-030-64144-3

Printed in the United States
by Baker & Taylor Publisher Services